Avkoanic View Microworld

By

Yecheskiel Zamir

© 2004 by Yecheskiel Zamir. All rights reserved.

Revised edition, August 2004, electronic books by AuthorHouse

No part of this book may be reproduced, stored in a retrieval system, or transmitted by any means, electronic, mechanical, photocopying, recording, or otherwise, without written permission from the author.

First published by AuthorHouse 10/11/04

ISBN: 1-4184-6513-5 (e-book)
ISBN: 1-4184-4573-8 (Paperback)
ISBN: 1-4184-4572-X (Hardcover)

This book is printed on acid free paper.

Dedication

To all individuals who attempt to search for the reality of the physical world, and to those who try to find the secret and evolvement of life from tiny particles, I dedicate this small book.

Author Statment

Early formation of matter and early creation of life are discussed in this book. Opening with a few questions makes it easier to deliver my message.
1. How far can matter split or be split?
2. Is there a tiniest particle in the universe?
3. What forces are *fundamental*, and how are they created?
4. Is there a *medium* in the cosmos, in which rays *can exist* and propagate astronomic distances *in all directions*?
5. What are the origins of life and of the DNA performance?

The Principle of the Tiniest Particle, termed **Avkoan Principle,** (Ch.1) echoed repeatedly in my mind as I pondered on these questions, until I wrote it down. That was in August 1973. Unwittingly, this principle was developed into a theory, termed **AVKOAN THEORY.** [1]

Indivisibility of the ATOM prevailed about 2450 years until 1913 when Rutherford found it composite (of a nucleus and electrons.) Ten years later, the nucleus was found composed of protons and neutrons (1923). Many discoveries of subatomic particles followed (1940-75). The quark model was raised with its surprising fractional electron-charge (1963). Presently, the **Avkoan Principle** has ended splitting of matter *forever*. This principle has lifted a banner declaring *a discovery of a particle smaller than the AVKOAN can never be expected because it can never exist*.

A very important conclusion is that the tiniest particle must consist of two *inseparable* intrinsic attributes: one *Avkoanic-mass* and one *Avkoanic-charge*. Even more important is the proof (Ch. 3) that mass and charge are *necessary and sufficient* for the creation and evolution of matter and the universe.

Ironically, pondering on this principle was an endeavor to answer the fourth question above, rather than the second. I was looking for a medium in the cosmos, other than ether, through which light rays can propagate astronomically *in all directions*, a medium in which rays *can exist* and propagate as *waves*, any kind of electromagnetic waves. The emphasis here is on the phrase *in all directions*. I published my results in a separate booklet entitled *The Cosmic Medium and the duality of Light Rays*. [2]

Avkoan Theory gave adequate answers to the opening five questions. The first and second were answered by the Principle of the Tiniest Particle. **Fundamental forces** in the third question are limited only to forces that are immediate interactions either among masses or among charges (Ch.4). They are created naturally either by **attraction** or **repulsion**. Attraction is satisfied by interactions among masses or among unlike charges. Repulsion is satisfied only by interaction of any two like charges, either both positive or both negatives. Shortly and solemnly, nobody would dare answer the fifth question. However, this book provides a bold and profound answer in Section 9.2.

The above discussion concerns only **Part I** of this book. Yet **Part II** reveals new perceptions of **heat, temperature, force** and **energy**. It answers the following two questions:

1. Do we really know what is **heat** and what is **temperature**?
2. Does **energy** conception precede **force** conception in the hierarchy of fundamental concepts?

Part II unveils a *limit* above which the concept *temperature* becomes **meaningless**. Above this limit we may speak only of *density*, i.e. quantity of material elements (or number of particle) per unit volume. Also this book made it clear that **energy** cannot exist without **fundamental forces**, which are immediately produced by interactions of the basic attributes of matter, mass and charge, i.e., by *attraction* and/or *repulsion* among masses or among charges.

After much hesitation I decided to include my poem, termed "Vampire," in this edition of my book. Its inclusion arose from the prevailing barbarity of human behavior worldwide.

Yecheskiel Zamir
October, 2003

[1] Y. Zamir: *Avkoan Theory, The Complete Volume I.* YZ Publications. Los Angeles. 1994.
[2] Y. Zamir: *The Cosmic medium and the Duality of Light Rays.* YZ Publications. Los Angeles, 1994.

© Copyright, January 1989 by Yecheskiel Zamir.

The following poem is copyrighted separately from the book. All rights reserved.

No part of this poem may be reproduced by any means or any form (including electronic, composing into musical notes, singing in public or inclusion in a movie production) without written permission from the poet/author.

Vampire

Vampire,

Just one Vampire,

And a dainty man,

Tranquil and naive,

Routinely occupied

In personal affair

Ever since,

Until ...

Abruptly

He's attacked,

Blood-sucked,

Flung down,

Snapped,

Toothed ...

Finally augmented

One more

Vampire.

And a prudent man,

Not a better repute his:

Self-secluded,

Isolated,

Entrenched,

As a rat...

But all of a sudden,

Through the narrow hatch,

He's reached,

In fraternity is set,

And in front of the limpid mirror

He too is a…

Vampire.

Vampire,

Just one vampire,

Only time went on

And everywhere roved

Swarms of…

Vampires.

As if …

Nothing is left but

A turmoil of…

Vampires.

Table of Contents

Dedication iii
Author Statment v
Vampire ix
Preface (To the first edition) xiii
Preface (To the second edition) xvi

Part One: BASIC PRINCIPLES OF THE AVKOAN THEORY 1

Chapter One: Fundamental attributes of matter 3
Chapter Two: Significance of the magnetic phenomenon 9
Chapter Three: Prerequisite conditions for the creation and evolution of matter. Necessity and sufficiency of mass and charge Elixirs 14
Chapter Four: Fundamental Fields of Forces, FFF 19
Chapter Five: The Law of Fundamental Forces (LFF) 22
Chapter Six: Impossibility of unifying Fundamental Forces, FF 26
Chapter Seven: The Bubble Model of FFF 29
Chapter Eight: Odor Analogy of FFF 35
Chapter Nine: Demonstrations of independence 38
Chapter Ten: Field of particles vs. field of forces 48
Chapter Eleven: Sources of singularities in the field equations 52
Chapter Twelve: Limitations of our mathematic systems 55

Part Two: NEW PERCEPTIONS IN THE MICROWORLD ... 59

Chapter Thirteen: On heat and temperature 63
Chapter Fourteen: On force and energy 86
Appendix A: The principle of the tiniest particle 96

Appendix B: A Marvelous Force 98
Endnotes 100
Bibliography 102

Preface
(To the first edition)

The complexity of particle behavior has been highlighted by successive novelties in the last few decades. It projects its influence onto several disciplines besides physics. Within this context, one may not treat the known theories as wooden absolute. To wit, "established theories" in many disciplines have increasingly become obsolete as later observations of reality are made with the introduction of leading edge technologies. The unknown reality brings many assumptions into questions underlying the present state of knowledge. It also burdens the scientific community with obligation to keep an open mind when propositions, challenging theoretical thoughts, are forwarded for discussion.

For example, the latest model of reality has established the *standard model* as the foundation of the structure of matter. Some forty years later experiments showed that quarks are not the smallest types of particles and are not the building blocks of the universe. Predictably, the smallest quark u is composite, since its mass is twice as big as that of the electron. Even the electron must be composite, as the scientific community accepted the existence of charges smaller than the electron charge, like those of quarks.

Part-I of this treatise is a revised version of the **BASIC PRINCIPLES OF THE AVKOAN THEORY.** The Avkoan Theory questions the assumption underlying the structure of matter apropos of exploring the existence of a tiniest particle termed **AVKOAN**. Twelve topics in this part are good enough for a first acquaintance with the theory. I extracted some of these topics from the original book, **Avkoan Theory, The Complete Volume I**, revised them and added the first and last chapters.

Other topics that are not immediately related to this subject have been ignored in this treatise. However, I want to mention here two chapters in the original book that introduced new ideas: Ch. 10 introducing *The System of Primary Particles*, and Ch. 11 introducing *The Electron Composition and its Configuration.*

A thorough reading of Part I would ease the comprehension of Part II, entitled ***NEW PERCEPTIONS IN THE MICROWORLD***. Part II introduces unfamiliar ideas of most familiar notions, viz., **heat, temperature, force** and **energy**. It comprises two chapters; the first (Ch. 13) presents a deeper insight of heat and temperature; the second (Ch. 14), makes comparison between force and energy. Alternative definitions are suggested to harmonize with **the principle of the tiniest particle** and eliminate confusion. The closing Appendix serves a handy reference to this principle, sparing the time to look up the original.

This book is primarily intended for college and junior high students. Its subject matter is mainly **particle behavior**. Teachers and scientists can benefit from its stunning revelations about heat, temperature, force and energy. Simple language is used to contribute to the understanding of unfamiliar ideas. I avoided using complicated Math but one: **The Law of Fundamental Forces, LFF**, which is indispensable for teachers and scientists. Skipping this would not impair the intended message of the book.

Yecheskiel Zamir
May 1999

Preface
(To the second edition)

AVKOANIC VIEW MICROWORLD is the second edition of the former title MICROWORLD. This change in the title is necessary to prevent confusing with several books entitled with the same word *Microworld*. "Avkoanic View" was an integral part of the original title. I dropped it just before printing for shortening. However, a sense of *chaos* may arise from the MICROWORLD conception. Restoring the "Avkoanic View" to the title will preserve the sturdy behavioral compliance of particles with fundamental laws of their own.

Grammatical and linguistic errors of the previous edition have been corrected in this edition. Chapter One opens now with five questions allowing the reader to ponder on, before he reads the stunning answers following them. Duplication of the DNA (in Sec. 9.2) has received now a deeper description. I avoided doing that in the first edition lest I slip in the genetic domain, which I reckoned out of the physical behavior of elementary particles. However, it was too late when I recognized that its inclusion is necessary to prevent ambiguity. Finally, a shorter chronicle of particle discoveries in Appendix-A replaces the original Appendix that appeared in the previous version of MICROWORLD. Appendix-B is a new Appendix, termed **A Marvelous Force** added in the present version.

Yecheskiel Zamir
July, 2004

Part One

BASIC PRINCIPLES OF THE AVKOAN THEORY

Yecheskiel Zamir

Chapter One

Fundamental attributes of matter

1. How far can matter split or be split?
2. Is there a tiniest particle in the universe?
3. What kinds of forces are *fundamental*, and how are they created?
4. Is there a *medium* in the cosmos, in which rays *can exist* and propagate astronomical distances equally *in all directions*?
5. What are the origins of life and the DNA performance?

As I pondered on these questions, The Principle of the Tiniest Particle, termed AVKOAN PRINCIPLE, echoed repeatedly in my mind, and reechoed incessantly until I wrote it down. It was unwittingly developed into a theory, termed AVKOAN THEORY.

Matter is composed of smaller and smaller particles. Scientists have always been intrigued about the smallest part of matter. Now we are intrigued by whether or not a tiniest particle exists. As a first step, the Greek philosophers coined the word **atom** (which implies *indivisibility*) when they speculated that matter is composed of indivisible tiny particles. Only in the 17th century did the THEORY OF ATOMS get experimental support by Robert Boyle and Isaac Newton. In 1802, Dalton's ATOMIC THEORY opened an era of discovery of many elements, before which they were unknown. The second step took place when the atom was found to be a composite, composed of a nucleus around which electrons revolve in a specific order. The atomic weight specified the type of element to which an atom belonged. In the third step, the nucleus of every atom was found composed of protons and neutrons. The number of protons specified the type of element to which an atom belonged. In the fourth step, several types of subatomic and subnuclear particles have been discovered. Most of them are much smaller than the proton/neutron. These discoveries have caused *repression* of thinking about a tiniest particle. A fifth step began with the introduction of the quark model. It is believed, but unproven yet, that every subnuclear particle is composed of one combination or another, one or more, of only six types of quarks that now are considered as the building blocks of matter and therefore of the entire universe. The type and number of quarks and their combining manner are believed to specify the type of subnuclear particle to which a particular quark combination belongs.

No cue has yet been found that might show a possible ending to this chain of probing/splitting matter into smaller and smaller parts, and no one can expect it to do so. Can anyone arrive at the ultimate (smallest) particle? It seems doubtful. However, a limit to this incidental procedure must exist and it must end at the ultimate (smallest) particle. This suggests a new step, the sixth, arising from an introduction of a principle termed **the principle of the tiniest particle** dubbed **Avkoan Principle**, stating the following:

> There must exist a group of primordial particles, termed *AVKOANS*, which are the tiniest, in a sense that there can never be expected a discovery of a smaller particle. Each *AVKOAN* necessarily possesses two intrinsic attributes, which are independent, indivisible, inseparable and indestructible: one **Avkoanic-mass** and one **Avkoanic-charge**. These are the **ultimate units** of mass and charge. Any particle whose mass or charge is bigger than the respective ultimate unit is divisible and cannot be tiniest; viz., it cannot be an *AVKOAN*. *AVKOANS* must therefore be identical in their masses and equivalent in the absolute value of their charges, each of which can be either positive or negative. Ergo, the entire universe is composed only of *AVKOANS*.

Note: I will show in Chapter 3 that no other fundamental attribute of matter is necessary for the creation and evolution of matter and the universe. In other words, mass and charge are necessary and sufficient to account for all phenomena and properties of matter known and unknown to man.

Based on the Avkoan Principle the following results are immediate:

1. Each AVKOAN generates simultaneously two fundamental fields of forces, **FFF**, gravitational, **G**, (due to its mass), and electromagnetic, **Em**, (due to its charge), Ch. 4 & 5.

2. We know that mass interacts only attractively with another mass or masses and so do unlike charges, while like charges interact only repulsively, according to law. (Formula 5.1, Ch. 5)

3. No interaction is possible between a mass and a charge, and their fields of forces do not interfere with each other's activity because of their independence. (Ch. 5 & 9)

4. Nature has given charges two exclusive virtues: **neutralization** and **repulsion**. Mass is exempted from both, but is provided with a **limitless accumulation** virtue into larger and larger masses. Fortunately, this virtue of mass is modulated by the two virtues of charges. [These provisions are necessary for the creation of matter out of primordial particles and for its evolution into complex material entities. Can anyone imagine what would exist without these provisions? It is enough to ponder about the amazing creation of the DNA in a variety of genes to get an idea how complex are the choices of particles to form into one type of entity rather than another.] (Ch. 3 & 9)

5. The Em-field of a particle is generated at such a high intensity that it overshadows the G-field of the particle entirely (Ch. 11), but this field (the Em-field) declines very sharply with distance and rapidly fades away within a few inches. Its rapid **repression** is caused mainly by interaction with nearby charges, Fig 7.2.

6. The G-field of a particle is much weaker than its Em-field at the beginning but declines very slowly with distance so that its influence persists for astronomical distances despite its asymptotic declination at large distances, Fig. 7.2. (Ch. 7 & 8)

7. The individual masses of a compound reinforce the G-field of one another, producing an intensified G-field. Every additional mass to a system/compound would intensify the common G-field according to the gravitational law. All masses in the entire universe produce a universal G-field that is common to all existing masses. On the other hand, only the net charge of a compound builds up the external Em-field of its unit. The internal Em-fields of all neutralized components in a unit become totally engaged in holding its components together. They impart none of their common Em-field to interact with external objects. Ultimately, these forces turn out to be the **binding forces** of the unit/compound. (Ch. 6, 8 & 10)

8. Mass does not reduce the intensity of the common G-field in a compound by internal engagement with adjacent masses, the way a charge does to its Em-field by neutralizing with a complementary charge. (Ch. 5 & 8)

9. Massless and chargeless particles are inconsistent. A neutral particle necessarily consists of at least two complementary charges neutralizing each other. (See Result 7 above)

10. No transformation between mass and energy is possible. The mass of every particle in a nuclear fission *carries* energy that agrees with Einstein's formula $E=mc^2$, but does not lose its identity. A light ray is a stream of particles each of which possesses a mass and a net charge, accompanied by Em waves. (Ch. 6 & 11)

Chapter Two

Significance of the magnetic phenomenon
The magnetic property is a "bi-by-product" of the charge

Electric and magnetic phenomena seem to depend on each other. However, the magnetic phenomenon is assumed to be a by-product of the **charge Elixir**[1] and totally dependent on it. Several reasons sustain this assumption, as follows:

1. We know that these two phenomena are related to each other: a moving charge always develops a magnetic field with two magnetic poles that accompany the originating charge wherever it may be, and a magnetic field always induces an electric current in a conductor that cuts through it. The assumption that a charge at rest does not develop a magnetic field is questionable, because of the relativistic character of the notion "at rest."

2. While opposite electric charges are easily separable; the magnetic phenomena always appear in a "dipole" condition, whether they are produced by a positive or a negative electric charge. No **magnetic-monopole-particle** has ever been

discovered, and it is extremely unlikely that a magnetic pole (say the N-pole) can ever be separated from its twin pole (the S-pole). It is conceivable, then, that a discovery of a **monopole particle** is impossible.

3. Being the tiniest existing particle, an AVKOAN can possess only one ultimate unit of each of its fundamental attributes (mass and charge) which, in principle, is independent of the other. Specifically, it cannot possess two units of the charge Elixir, and it cannot change its charge from positive to negative, and vice versa. The AVKOAN can possess either a positive or a negative Avkoanic-charge. This charge is not only unchangeable but also unexchangeable with a corresponding Avkoanic-charge of another Avkoan (unaccompanied by its connate mass). Obviously, an AVKOAN cannot have both positive and negative charges. The same principle must apply to a *magnetic monopole*, if it is considered an independent quality of matter.

4. Now, due to its charge, a spinning/moving AVKOAN develops both opposite magnetic poles, which are inseparable from the AVKOAN and from each other. Analogously, an imaginary spinning "monopole Avkoan" should produce two opposite Avkoanic charges in the same AVKOAN, which are inseparable from the AVKOAN and from each other. In other words, the existence of one of these two qualities (electric

charge and magnetic pole) implies the existence of a <u>pair</u> of the other in the same primordial particle. An AVKOAN therefore cannot possess both qualities as Elixirs. Besides its mass, it may possess either a charge or a magnetic pole as an Elixir. Which of these two is qualifiable? The choice can be in favor only of the charge. This leaves the magnetic property as the dependent one.

5. A particle having both opposite magnetic poles (N & S) cannot be primordial/tiniest unless the magnetic property is dependent, because the tiniest particle cannot possess more than one Avkoanic unit of the same Elixir. If for any reason the **monopole doctrine** persists and the magnetic property is considered independent, then the Avkoanic charge necessarily becomes a by-product of the "assumed magnetic Elixir." Alternatively here, two opposite Avkoanic charges must develop by the spinning of the *proposed/assumed monopole particle*, and the electric induction phenomenon among AVKOANS would then become a directional rearrangement of their bi-electric charges. Such a situation is improbable because the tiniest particle cannot have a pair of charges. This leaves the magnetic property again as the dependent one, and a further discussion becomes superfluous.

6. All attempts to detect an isolated *magnetic monopole* in the last few decades have been unsuccessful. In 1982, Blas

Cabrera of Stanford University announced that he detected the presence of a magnetic monopole experimentally at a very low temperature (about $10°K$ or $-263°$ C). He concluded this from a sudden jump in the current of a loop made of a superconductor metal, and considered it caused by the passage of a magnetic monopole through the loop[2]. A single experimental observation cannot serve as evidence, although it may, as it really did, inspire a hope for further experimental sustenance from other sources. Yet this event was never repeated. No other sources have yet sustained it and, according to The AVKOAN THEORY, none ever will.

In any case, the debate about the *magnetic monopole particle* cannot refute the strong conclusion of The AVKOAN THEORY that the magnetic property is a by-product, more precisely a **bi-by-product**[3] of the charge Elixir and totally dependent on it. If this were not so, then the charge would have to be a by-product and dependent on the *proposed magnetic monopole*, assuming this is an Elixir, which is impossible.

What is intriguing in this regard is not the success or failure of recurring attempts to observe a ***monopole particle***, but how the **monopole doctrine** could impose a successful illusion. A profound discussion on the magnetic monopole is given by Heinz Pagels in his remarkable book Perfect Symmetry, (Chapter 2.)

We may conclude that the magnetic property is dependent with no chance of becoming an Elixir. It is only a "**bi-by-product**" of the charge Elixir. Since it is always developed in <u>pairs</u> of magnetic poles, a *magnetic monopole particle* has no physical reality. This conclusion is made possible by the virtue and integrity of the notion "tiniest" characterizing the Avkoan, that leave no possibility for the Avkoan to possess a pair of the same type of Elixir.

True, many particles other than AVKOANS develop two opposite charges. This becomes possible by the fact that these particles are composite. They already contain both charges and only a rearrangement in the mutual locations of their components takes place.

Chapter Three

Prerequisite conditions for the creation and evolution of matter. Necessity and sufficiency of mass and charge Elixirs

In his humorist wit, Einstein proposed that God did not play dice when He created the world and God had no choice in *adjusting* the parameters (such as mass) to make different universes. The following analysis finds no necessity for God to play dice, once He adjusted the ultimate units of mass and charge. These are the only two attributes of the tiniest/primordial particle. Mass and charge are necessary and sufficient for the creation and evolution of matter and the universe, independently of temperature as are described ahead. (Also see Sec. 13.3 & 13.4 concerning temperature.)

While **mass** is necessary as the essence of matter, **charges** are necessary to help masses accumulate and aggregate into discrete units of matter. Unless each primordial particle possesses both mass and charge, none of them could probably form into a material object. Moreover, only both *attraction* and *repulsion* processes of these Elixirs can create matter. The fields generated by masses and those generated by unlike charges satisfy attraction, while the fields of like charges satisfy repulsion.

If the tiniest particle were to possess only a mass Elixir without a charge, then this mass would generate only a gravitational field **G** and interact attractively with other masses. So, the entire universe would finally become an indivisible, unchangeable, and everlasting gigantic mass.

If the tiniest particle were to possess only a charge with no mass, then each charge would originate only an electromagnetic field of forces **Em** that might interact with others without possible formation of matter. We cannot surmise how these charges would rove about without something that "carries" them. A still worse situation would be an existence of only like charges (whether all male or all female[4]). Then the entire universe would be composed of primordial charges continually repelling one another, expanding into space forever, not as material particles, but merely as charges. Anyway, matter could never be created then.

If both male and female charges were to exist without masses, then every complementary pair of charges would become neutralized into a zero net charge. The combined pair could probably suspend or rove about in space without ever being able to accumulate and combine with others to form into a material particle. We cannot imagine what form such a pair of charges would take without mass, since these charges are Elixirs and are therefore indestructible.

If each and every primordial particle were to possess ultimate unit of mass and ultimate unit of charge and all charges are male or all of them female, then none of these particles can combine with another. The universe would be composed of primordial particles distributed uniformly throughout space, equidistant from adjacent particles (whether all male or all female), so that the gravitational attractive forces of each particle will balance its repelling electromagnetic forces with those of its neighboring particles. The universe would be neither expanding nor contracting, and probably no dramatic motion would take place. This situation is very important. It does exist in the deep space where Negakoans (female AVKOANS) prevail and maintain equidistant separations from adjacent Negakoans so that space is uniformly distributed with Negakoans, making the space negatively charged.[5]

A situation of primordial particles possessing *ultimate* units of unequal intensities of either or both Elixirs is out of our discussion. It is heraldic in The AVKOAN THEORY that any particle possessing a mass or a charge larger than those possessed by the Avkoan is divisible and cannot be tiniest. All primordial particles must be identical in their masses and equivalent in the absolute value of their charges, which can be either male or female in every AVKOAN. Moreover, each AVKOAN must possess an ultimate unit of mass and an ultimate unit of charge.

Thus, the necessity of both Elixirs and both complementary charges is satisfied. *Masses can accumulate into matter only by the help of attraction and repulsion carried out by their connate charges.* The combined activity of masses and charges satisfies the sufficiency as well, not only for the creation of matter but also for the evolution of matter and the universe.

The relativity of motion and rest is included in the definition of Fundamental Force (**FF**). Since attraction and repulsion can occur only among masses or among charges (no matter what object may pass between them) neither a fixed universal frame of reference nor a moving one is necessary. All particles and objects are related to each other through their mutual interactions.

Creation: Charges develop electromagnetic fields of forces only when they are in motion, a fact that became the tool by which matter is created from primordial particles. Thanks to their masses, these particles could accumulate into discrete **units** of larger particles. In turn, they combine with similar and dissimilar units to form still larger **units** of different substances at different levels/generations, starting from the Avkoanic generation and evolving into the quark, nuclear, chemical, planetary, galactic and intergalactic generations. *Each unit in every generation is well defined not only in mass and charge but also in size, relative position, and separating distances* from its neighboring units within a larger unit of a higher generation. Each unit is destined to become a part of a more advanced unit in a higher stage of evolution. (Sec. 13.10). Every particle plays its individual role in the foundation of the universe, and so every material object does. And each of them is part of the overall construction and evolution of the universe. They are all *mutually responsible* to each other's behavior and to their collective relationship with one another. Amazingly, everything in the universe owes its existence to a **dual Elixirian activity: Mutual attraction among some AVKOANS, enhanced by repulsive force with**

others, according to law, (Formula 5.1). This activity is true in the macroworld as well. It is only necessary to generalize the notion AVKOANS into the notion "particles" or "material objects." It is also inherited into all material entities including living creatures and human societies. Repulsive forces amoung individuals or groups, enhanced by attractive forces with others, have brought into combative arguments that often erupt into bitter fight and bloody wars.

Chapter Four

Fundamental Fields of Forces, FFF

Each Avkoanic-Elixir generates, independently and simultaneously, its characteristic field of forces: The mass Elixir generates a **gravitational field (G)**, while the charge Elixir generates an **electromagnetic field (Em)**. Both fields are inseparable from their initiator, the Avkoan, wherever it may be, and whether it is at a relative motion or rest. These are the two **Basic Fields** of an AVKOAN and are therefore the only two Fundamental Fields of Forces, **FFF,** in the universe. Only at a relative motion/spinning does the Avkoanic "electric" field produce a magnetic field on its own. In actuality, this (magnetic) field is a *sub-field* that intertwines with the mother (electric) field into a unified **electromagnetic field (Em)**, as is normally produced by every moving charge. The electric field has only one center of generation, coinciding with the Avkoan's center. On the other hand, the magnetic field has two centers (twin poles), neither of which coincides with the center of the AVKOAN (which is also the center of the electric field). Each side of this pair of magnetic poles has a **magnetic-gender** distinct from the other side in its behavior. Repulsive fields of forces are developed among similar magnetic poles, and attractive fields of forces between "opposite" magnetic poles, in exact analogy with the mutual behavior of charges.

The spinning rate or the translatory speed of an AVKOAN determines two important magnetic behaviors: 1. The intensity of the magnetic poles. 2. Exact phase in which the electric and the magnetic fields interlace. The ideal situation is achieved at the speed of light in a translatory motion placing these two fields at right angles to each other.

Each Elixir of an AVKOAN conforms to its relative law of forces: Its mass Elixir conforms to the gravitational law of attractive forces among masses (to agree with Newton's gravitational law). Its charge Elixir conforms to the laws of electric and electromagnetic fields of forces according to Coulomb's law and Maxwell's equations. Attractive forces are thus developed between oppositely charged AVKOANS and repulsive forces between identically charged AVKOANS. Magnetic poles interact in a similar way to charges:

$$\text{(gravitational)} \quad F_g = G \frac{m_1 m_2}{r^2} \quad (4.1)$$

$$\text{(electric)} \quad F_q = \pm k \frac{q_1 q_2}{r^2} \quad (4.2)$$

$$\text{(magnetic)} \quad F_p = \pm \mu \frac{p_1 p_2}{r^2} \quad (4.3)$$

In these Equations, F stands for force, m for mass, q for charge, p for a magnetic pole, r for distance, G, k, and μ stand for proportionality constants.

Unlike an electric charge, mass is in no way considered a negative quantity. Moreover, though mass is always considered positive, the forces developed among Avkoanic masses can only be attractive forces. This is in direct contradiction to forces developed between identical charges. The reason for this contradiction is still an enigma. It is taken for granted to agree with observations, as far as entities within the Milky Way Galaxy, and our conceivable universe, are concerned. This fact influences two important phenomena:

a. Establishment of the specific properties of subatomic particles through the Avkoanic Elixirs.

b. Equidistant locations of the AVKOANS in deep space.

Avkoan Theory does not discuss nor comment on what is called antimatter.

Chapter Five

The Law of Fundamental Forces (LFF)

Fundamental Forces, **FFs**, have no self-existence. Either attraction or repulsion between at least two interacting Elixirs of the same quality creates them: either two masses or two charges. [Magnetic poles are proven by-products of electric charges (Ch. 2) and they are used here only for comparison]. These forces of nature have been given remarkable mathematical expressions by two different laws, Newton's Law of Gravitation and Coulomb's Law of Electric and Magnetic forces, despite their extensions according to the general theory of relativity. More about this extension will come in Ch 12. Both laws follow the same procedure for creating fundamental forces, and they are **unifiable** into a common law, (Formula 5.1), which may be called the **Law of Fundamental Forces (LFF):**

First, mass interacts only *gravitationally* with another mass or masses, whereas charge interacts only *electromagnetically* with another charge or charges. **No interaction is possible between a mass and a charge.** Masses interact only attractively with one another, and so do unlike charges. Like charges interact only repulsively, so that a force is always developed between any two interacting Elixirs of the same quality. [For the sake of comparison, this may be added: a magnetic pole interacts only electromagnetically with another magnetic pole. No interaction can be between a mass and a magnetic pole. Unlike poles attract while like poles repel.]

Second, a fundamental force is proportional to the product of magnitudes of the interacting Elixirs and to the reciprocal of the distance square between them, Formula 5.1.

$$F = \pm \gamma \int \frac{x_1 x_2}{r^2} dr \qquad (5.1)$$

$$\gamma = G \qquad \text{(for gravitational force)} \qquad (5.1a)$$

$$\gamma = k = \frac{1}{2\pi\varepsilon} \qquad \text{(for electric force)} \qquad (5.1b)$$

$$\text{(effectively)} \quad \gamma = \mu = \frac{1}{2\pi\mu_o} \quad \text{(for magnetic force)} \quad (5.1c)$$

In these equations, **F** stands for the force between two interacting-Elixirs (or magnetic poles), x_1 and x_2 stand for their respective magnitudes; r for the distance between them and γ for proportionality constant. It replaces G in a gravitational interaction, k in an electrical interaction, or μ in a magnetic interaction; ε stands for dielectric constant, μ for permeability and μ_o for magnetic induction in "empty" space. (Cf. Ch. 4)

Unless an interaction or a chemical reaction takes place, the quantities x_1 and x_2 retain their fixed values (except in magnetic poles), while the distance r between them undergoes continuous change until it attains its limit.[6]

The constants G, k, ε, μ and $μ_o$ in formula 5.1 and related equations are not absolute. They are **fluctuant** approximations depending on the regional density of matter. Being the only two attributes of matter, mass and charge are the only two **absolute constants** in the universe. Even the velocity of light cannot retain an absolute constant value. It fluctuates below and above its currently popular value when it approaches or passes spaces of different densities of matter. The greater changes in density of the medium intervening between interacting Elixirs, the greater the proportionality constant deviates from its normal value in a *particular* force. Due to the long range of the influence of mass, the constant G appears as undergoing more fluctuation than k and μ, although it is the most stable constant following the absolute constancy of mass and charge. **The fluctuation in the value of G can serve as a good measuring tool for the distribution of mass in the cosmos**, and particularly for the relative locations of what are termed "black holes" in space. On the other hand, the short range of the electric field appears to imply fewer fluctuations in the values of the constants k and μ. Shortness in ranges of the electric and the electromagnetic fields is a consequence of neutralization between unlike charges. Validity of each of these proportional constants in our galaxy does not automatically guarantee its validity in other regions of the universe where the laws of nature apply. Each of them depends on the density of the medium that, in turn, depends not only on "temperature" but also on several other factors that are not known to us right now. No difficulty is expected in grasping the fluctuating values of the acceleration g of falling bodies on earth. Yet it is not that easy to grasp on earth the fluctuation of the universal gravitational constant, G. Perhaps a profound and deep analysis may be required for that.

Chapter Six

Impossibility of unifying Fundamental Forces, FF

Much pain has been taken in the attempts to unify all the assumable existing "four" *fundamental* forces (considering weak and strong interactions as *fundamental* forces) before the announcement of the possible existence of a "**fifth**" type of "fundamental" force.[7] Several methods have been devised to relate these forces to one another. A surmise has been adopted that they do so at specific fantastic temperature-thresholds where particles are assumed to undergo similar modes of interactions. A smart procedure has been surmised to explain the creation of nuclear forces envisaged "**particle exchange**" within the nucleus of every atom. Imaginary particles termed "gluons" and "gravitons" and a new theory termed "Gauge Theory" have been necessary to make possible a unification process termed GUT (Grand Unified Theory), which has finally been accepted. Suddenly, the GUT has been violently confronted with a new agonizing problem elicited by the possible existence of a **fifth** type of force, so that the whole edifice of the GUT began to collapse. A startling *turmoil* of several simulated forces is expected to emerge out of illusiveness and will repeatedly surprise us and disrupt the excellence of our modern knowledge whenever new particles smaller that those already known to us are discovered. The smaller the particle, the shorter is the attainable separation among the components and therefore the stronger is the force, holding the components. This force demands its specific rank in the turmoil of misleading *duplicate* types of forces. The adoption of *weak* and *strong* interactions has inevitably triggered a source of *recurring* problems issuing from surmising new types of forces claiming their turn in the unification fantasy. They demand desperate attempts repeatedly to give them full satisfaction, unless we *renounce* the idea of the possible existence of fundamental forces other than **G** and **Em**.

The definition of fundamental forces (given in Ch. 5) disproves the existence of fundamental forces other than G and Em. Therefore, the so-called GUT cannot unify, if ever, more than these only two existing fundamental forces. Surprisingly, all the queer *ritual* in honor of the unification attempts has no justifying basis apropos to reality. **No unification whatever can be considered between the two Basic Fields of forces G and Em.** In vain have scientists concentrated their efforts on devising and imposing unification which *a priori* was doomed to failure. (See Ch. 7 & 11). It is like imposing marriage on an unmatched couple and expecting them also to unify their veins, nervous systems and all other organs together into one unified body by some peculiar molding or fusing process.

What is intriguing now is not the unification itself of "four" or "five" forces, but we are required to find out why the two Basic Fields themselves appear **unifiable**. The lure for unification must have been so strong that the elite of our scientists strove to knit the mesh into which they themselves were destined to fall, as they did. Eventually, they got entangled severely enough to waste much time and effort in the attempt to solve many superfluous problems emerging from the unification attempts. The impossibility of unifying fundamental forces arises from the following: First, unification is irrelevant for forces other than **G** and **Em** simply because they are non-fundamental. Second, it is inapplicable for G and Em either, because of the very fact that they assume **independent activities** inherited from the independence between mass and charge. Being Elixirs, mass and charge assume independent activities, although every Avkoanic mass is eternally associated with its connate (inborn) Avkoanic charge (See Ch. 9.) As mentioned in Ch. 5, interaction between a mass and a charge is impossible. The gravitational field of a particle is absolutely independent of the particle's own Em field. Accordingly, neither interaction nor unification can take place between their fields of forces. Repeat, no unification is possible between their fields at all. The G field may react (rather than interact) with the Em field of *another* particle in either supporting or impeding a manner. This type of reaction is not a fundamental interaction because it is not an immediate process of attraction or repulsion between masses or between charges. A **G-field** can only be produced by interactions between two or more masses, whereas an **Em-field** only between two or more charges.

Chapter Seven

The Bubble Model of FFF

7.1 Ranges of G & Em of a particle

Both mass and charge of a particle (or a system of material objects) generate their respective fields of forces simultaneously and independently of each other. The more Avkoanic masses an object contains the denser flux its combined **G** field has. Its **Em** field develops a flux as dense as its net charge can provide and it extends no more than this net charge can afford. However, the **G** field of a particle is not everywhere dominant. The distance from the source of these fields plays an important role. Both fields are weakened with distance. However, within a short critical distance the flux density of the **Em** field of a system is much higher than that of its **G** field. So the **G** field has little or no practical effect on its immediate vicinity. Beyond another critical distance the resultant **Em**-field of a system fades away entirely, leaving a dominance to the G-field. The following model shows these relationships clearly.

7.2 The Bubble Model of an isolated particle

Imagine two concentric spherical bubbles assigned to every particle, having a common center at the particle's center. Let r be the radius of the **Inner Bubble, IB,** and let R be the radius of the **Outer**

Bubble, OB, as shown in Fig. 7.1. These two bubbles represent two boundaries of the influential extent of the two **Basic Fields** of a particle. At a range of distances within the **inner bubble, IB,** the flux density of the **Em** field of a particle is very high. Its Em field overshadows the influence of its **G** field, giving an impression that the particle has only an **Em** field but no **G** field at all, despite our certainty of its presence. Practically, the overshadowing of the G field by the Em-field, within **IB**, is strong enough to render the G field negligible and undetectable. At distances longer than r but shorter than R, the G field appears as well and simultaneous operations of both fields are observed.

At distances longer than R, the intensity of the Em field becomes too *feeble* to be discerned or detected. It is then practically negligible,

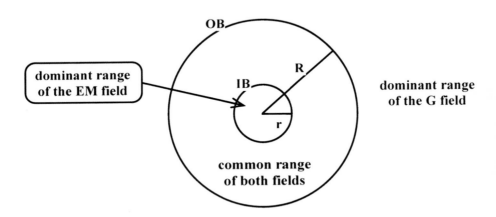

Figure 7.1: The Bubble Model of G and Em fields of a particle. R is the radius of the outer bubble OB and r is the radius of the inner bubble IB.

making an impression that the particle has only a G field but no Em field at all. The G field becomes dominant beyond the **OB** although it continues its declination the further it is from the originating particle, but no trace remains of the Em field.

This Em field at first very strongly *overshadows* the G field but declines sharply until it fades away at the **OB**, while the G field persists, although it continues to decline very slowly. (Fig. 7.2)

The radii of the two bubbles are different for different particles. Both bubbles can be smaller for one particle and larger for another, depending on the net charge of each particle and on the permeability of the surrounding medium even for particles of equal masses.

The radii R and r depend only on the intensities of both mass and charge of the particle. The ratio R: r depends not only on the ratio

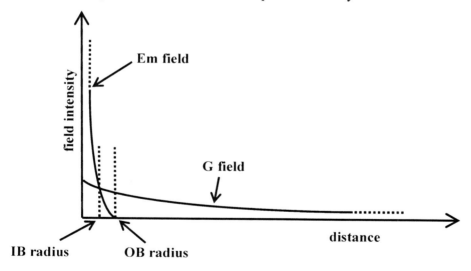

Figure 7.2: The intensities of G and Em fields of a particle as functions of the distance

e:m, but also on the volume of the particle, and it is determined mainly by the complex attractive and repulsive forces among the components.

Note: Throughout the rest of this treatise, any discussion about a particle is also applicable for a system of particles. Similarly, any discussion about an object is applicable for a system of objects, and a discussion about an Elixir is applicable for an **ultimate** unit and an **accumulative** unit or a system of Elixirs of the same quality.

7.3 The Bubble Model of two particles

Fig. 7.3 illustrates two particles A & B. The symbols r_1 & r_2 denote their respective **IB** radii, while R_1 and R_2 denote their respective **OB** radii. Let d be the sum of both **IB**-radii (Eq. 7.1) and D the sum of both **OB** radii (Eq. 7.2)

$$d = r_1 + r_2 \quad \quad (7.1)$$
$$D = R_1 + R_2 \quad \quad (7.2)$$

Within a separating distance d, the resultant **Em** field of the particles A & B is dominant and no discernable gravitational interaction is noticed between them. At a distance beyond D the intensity of their common **G** field becomes dominant and no discernable **Em** interaction takes place between them. The farther the particle A from B, beyond D, the weaker becomes their common **G**-field. At a distance x between d and D, both fields are observed operating together, but their respective effectiveness changes according to the length/span of this distance. Any object within the dominant field of another is affected accordingly and it responds accordingly.

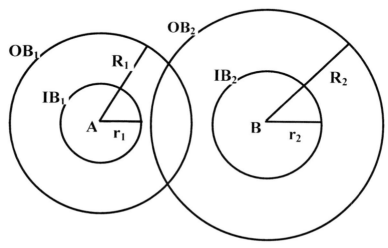

Figure 7.3: The Bubble Model of G and Em fields of two material entities A&B

Consider an object B approaching another object A. How would the fields of the object A influence the object B? When the object B is still outside the Outer Bubble **OB** of A, it is affected only by the G field of A. Between **OB** and **IB** of A, it is affected by both fields of A. Within the **IB** of A, the object B is affected only by the Em field of A. Here, the object B encounters no discernible effect from the G field of A. It interacts only electromagnetically with the object A. Meanwhile, similar effects are imposed by the fields of the object B on the object A.

When the distance between A and B is equal to or less than d, the objects are influenced by each other's Em field and no discernible effect is shown by their G fields. The closer A and B are to each other within this critical distance d, the stronger they are influenced by each other's Em field. At a distance x between d and D, only the G fields of each of the objects A and B will influence the alternate object, but both of them retain their individual Em influence on objects that interpose between them. At distances greater than D, both of their individual **Em fields** fade away entirely, leaving only their individual **G-fields** attractively influential on each other's source. The further A and B become from each other beyond D, the less they are affected even by each other's G field.

Similar to **complementary charges**, the *electromagnetic* fields of several particles of **complementary gender** *neutralize* each other, allowing only their **resultant Em field** (i.e., what remains of these fields after excluding all *neutralized* fields) to interact with external objects. Therefore, every point in space is definitely assignable whether dominated by the **G** or the **Em** field of a particle/substance, depending on the following factors:

1. *Distance from the particle generating these fields,*
2. *Mass size of the generating particle,*
3. *Charge size of the generating particle,*
4. *Medium permeability.*

Chapter Eight

Odor Analogy of FFF

Independence between **G** and **Em** fields becomes comprehensible by an analogy with a botanic field of a certain plant in which the individual plant consists of only one odorous green leaf embracing only one odorous red flower. The flower's odor is quite distinct from the leaf's odor. In certain weather conditions such as temperature, pressure, rain, humidity, wind, seasons, insects, etc., the plant grows a leaf without a flower. The flower's odor will be missing around that region, leaving an impression that the plant consists only of a leaf and grows no flower at all, as can be judged by the leaf's smell and confirmed by its green color. Under other conditions a flower grows without a leaf, leaving an impression that the plant consists only of a flower and grows no leaf at all, as can be judged by the flower's smell and confirmed by its red color.

One could easily be tempted to believe that both types of odors could be the same, although in a dry season, for instance, it smells differently than it does in a humid season accompanied by a change in color. A fault could also be in our sensitivity of smell and color under different weather conditions. One would then be inclined to believe that both types of odors could possibly *unite* into one original type of odor characteristic not only of this specific plant but also of all plants, no matter what color they show. A support to this possibility would serve the normal condition apropos to simultaneous growth of a leaf and a flower. The situation with **G** and **Em** fields in particles is essentially similar to this fictitious story.

Every particle and so every material entity generates both its G & Em fields simultaneously. In some occasions however, one of these two fields becomes undetectable. This situation depends on the magnitudes of its mass and its net charge. A neutral particle for instance, shows no Em activity, making an impression as chargeless, having only a mass. In actuality it has at least two complementary charges that neutralize each other. On the other hand, a particle with intensive net charge originates such a strong Em field that it overshadows its G field almost entirely, making an impression that it is massless and has only a charge. Although the neutrino was known as a neutral particle, it was debated repeatedly as being massless, ignoring my persistent disapproval since 1985.

The mathematical field-equations have prompted some temptations to unite both fields mathematically, but none of these attempts was successful. Reasons for this failure can be clearly understood from Ch. 6, and by musing through *The Bubble Model* (Ch. 7), *Odor Analogy* in this chapter, and *the sources of singularities in the field equations* (Ch. 11.)

We know that interactions among masses do not involve neutralization as an interaction between complementary charges does. **The intensity of the G field of a particle can never be annulled or weakened by internal interactions among masses, the way the Em**

field is annulled/weakened by a neutralization process. True, the intensity of the **G field** is controllable by annexing or reducing a measurable amount of mass. Annexation of a mass however, is inevitably accompanied by a corresponding annexation of its connate (built-in) charge. The same is true for the reduction process of mass. Any change in the G field of a material entity is necessarily accompanied by a corresponding change in its Em field, thus sterilizing our controllability regarding the G field.

On the other hand, **complementary charges** (balancing amounts of male and female charges) neutralize each other. A combined pair of particles stores all, or most of, the Em-forces between the pair. So, a neutral particle shows no external Em activity except perhaps a *feeble* magnetic field due to spinning, vibration, or any other type of relative motion. The neutralization property inspires possible experimental evidence for the independence of G and Em, as will be shown ahead, in Ch. 9.

Chapter Nine

Demonstrations of independence

The Bubble Model in Chapter 7 describes the independence of G and Em fields of a particle and their relationship in those of two material entities. Since the Avkoanic-charge is inseparable from its connate Avkoanic-mass in each Avkoan, isolating a collection of masses without their inborn charges, or a collection of charges without their inborn masses, is impossible. This fact undermines most experimental attempts to confirm the independence of G and Em.

Theoretically, depriving one of two interacting masses of its gravitational property invalidates the gravitational interaction between them, as if the other mass, too, is deprived of its gravitational property. However, no change occurs in the intensity of the resultant Em field of their inborn charges. Similarly, depriving one of two interacting charges of its Em activity causes an automatic disappearance of their resultant Em field, but no change occurs in the intensity of the common G field of their connate masses. Nevertheless, independence of the two Basic Fields is remarkably expressed by examples as described ahead:

1. By Millikan's experiment in the particle domain.
2. By the **DNA** reproduction process in the life domain.

9.1 Millikan's oil drop experiment

Millikan's experiment shows the independence of G and Em fields vividly. Millikan apparatus, Fig. 9.1, consists of two metallic plates assembled horizontally parallel to each other and to the surface of the earth. One plate is connected to the positive terminal and the other to the negative terminal, of a D.C. electrical power through a potentiometer that controls the intensity of the Em field across the plates. The flux density of the Em field is proportional to the voltage across the plates and depends on the distance between the plates, which is adjustable mechanically. The plates are maintained at a fixed chosen distance apart during each consecutive experiment. No matter how we change the intensity of the Em field, no change occurs in the intensity of the G field between the plates. The G field maintains its original intensity throughout these experiments exactly as it does when no voltage is applied across the plates.

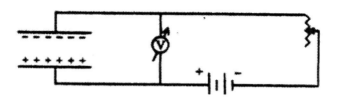

Figure 9.1: Millikan's circuit

All the results obtained from experiments using Millikan apparatus have used the acceleration g of falling bodies (taking into account its specific values in different geographic locations on the earth.) If the Em field between the plates *does* affect the intensity of the G field, then all the results of the said experiments lose their credibility. However, this cannot be so, because these results were not disproved by other experiments.

9.2 Origin of DNA and power of independence

Origin of DNA: Life is a baffling secret of Nature. It will always be so until someone reveals this secret. Scientists could not comprehend how life began. Biologists may consider DNA as the origin of life on earth and the existence of living creatures as made possible by duplication and reproduction of **DNA** molecules. Speculations may suggest that life started with the *appearance* of tiny primitive creatures and duplicated just as lifeless* substances did. Some lifeless material managed to acquire properties of *organic molecules* in the long run. Some organic molecules could have transformed into tiny, quasi-living creatures, and gradually evolved into tiny primitive living creatures. These tiny creatures evolved further into a variety of plants and animals. But *how* the initial tiny creatures appeared on earth in the first place? Scientifically, nobody knows. This book does not discuss unscientific belief.

Prior the beginning of life on earth only lifeless* matter could exist. Duplication of lifeless* substances grew naturally much as a crystal could grow in saturated solution. These lifeless substances evolved into primitive living creatures during billions of years. Yet, nobody knows *how*. Some of these creatures reproduced in a huge number. Others have been extinct eventually until the appearance of primitive animals, which evolved into advanced and more advanced animals having higher level of DNA complexity.

*   **Note:** Here the concept "lifeless" refers to material that is not organic, i.e. excluding organic tissues of dead animals and plants.

Avkoan Principle opened a new avenue explaining the **original** reproductive power of *material* entities. The existence of this power is an immediate result of the basic attributes of matter, mass and charge, and their basic fields of forces *gravitational* **G** and *electromagnetic* **Em**, without which no DNA could exist. These attributes of matter are delineated in the **Principle of the Tiniest Particle** and proven *necessary and sufficient* for the creation and evolution of matter and the universe (Chapter 3.) Duplication of matter is based on normal fundamental behavior of mass and charge, thanks to their two basic fields-of-forces *gravitation* and *electromagnetic*, i.e., **attraction** and/or **repulsion** either between masses or between charges. These are the only forces that impose behavioral rules on particles and subsequently on matter.

In early living creatures, duplication and reproduction began with absorption of particles to meet basic physical and chemical, or (phys-chemical) rules. Lifeless molecules absorb *raw* material, not necessarily involved with organic material. Generative cells absorb *ripe* organic nutrients in compact molecules, and living creatures require *organic* molecules.

Egg or womb necessity: Reproduction underwent many progressive improvements, harmoniously with periodic evolution of living creatures, until it turned into DNA. Advanced reproduction has gradually overshadowed the phys-chemical process. Duplication of the DNA has become ever more complex by evolution. In advanced animals, it became complex enough to require specific warm conditions for growing and surviving within an *egg* or a *womb* to produce only a few fragile babies, mostly only one at a time. Within a womb a fertilized egg gets its nutrients from the mother blood through a specific mechanism that connects the egg to the womb interior walls throughout pregnancy. It is protected there within a specific reproduction organ of the animal against exterior particles and bacteria. Nutrient supply is necessary to feed the fetus through the mother blood. After being born, the infant still requires to feed on its mother milk for a few months. The more an infant is fed on its own mother milk, the better it grows. The longer it does before weaning, the healthier it becomes.

In species that grow in water without a womb, the egg gets nutrients found in water, per osmosis process. So it is with reptiles that bury their eggs in damp surfaces. How does a bird egg get its nutrients? It only gets warmth from the mother or father body by covering the egg during a few weeks. How does the egg transform its contents into a living creature before it hatches? It is the duty of genetic explorers to reveal this mystery of nature.

However, the original absorption process did not disappear, because it is a fundamental property of particles. It did not and will never disappear despite apparent domination of complex (organic) duplication of the DNA in the advance animal kingdom. Without the former (i.e., without the phys-chemical process) no organic process would ever exist either. Recall the Avkoan Principle provides matter as being composed of a **mass** and a **net charge** that produce a gravitational force **G** and an electromagnetic force **Em**. We are amazed by recent discoveries of the DNA performance. Absorption of large molecules, well prepared for assimilation, has replaced absorption of crude *raw* material.

The DNA performance is *abnormal*, because it is not a normal performance of material particles. Normal performance of particles is the immediate interactions of masses or of charges, which are the activities of their fields of forces **G** and **Em**. The DNA performance may be called *supernatural* performance of living matter. It requires *abnormal* conditions available in living body or molecule in specific temperature range. We must admit that our planet itself is *supernatural*. It has *abnormal* conditions that ease creation, survival and continuation of life on Earth. A radical change in just one climate item would be enough to destroy all living creatures on Earth. Superiority or abnormality of planet Earth is considered unique and not found elsewhere in the entire universe.

In short, advanced assimilation of DNA is not the only existing reproduction way of the DNA. The phys-chemical absorption process of duplication would not disappear because it is a primary activity of matter. How duplication evolved into complex DNA is still an intriguing topic.

Biochemists have recently created some necessary conditions to fertilize animal egg and even human egg in a test tube. However, it is extremely doubtful that all the required conditions are achievable outside a womb. A clone mechanism would never procreate a healthy living creature without protection within a womb. Womb protection is necessary for breeding a healthy offspring. Without protection of a womb the fetus must lack something unexpected but vital. It may excel in some qualities of normal breeding but lack other vital qualities that leave it defective or monstrous. Cloning would therefore produce disaster on humans.

Power of independence: I propose that life on earth owe its existence to the *independence* of G and Em. (Millikan's oil-drop Exp., Sec. 9.1) In principle, reproduction process is carried out by discrimination through the *independence* property of G and Em as follows. Due to its mass, a particle is attracted gravitationally to a massive molecule when the particle happens to be within an effective **G range** of the massive (absorbing) molecule. (Sec 7.3 & Figure 7.2) Unless the particle reaches an appropriate distance within the influential **Em range** of the massive molecule, it will not combine with the molecule. The particle will not join the molecule in spite of its gravitational attraction to the molecule. So, no assimilation or reproduction would take place by the molecule (Bubble Model, Ch.7). Assimilation and reproduction would occur when the particle comes within the influential **Em range** of the absorbing molecule, taking into consideration other factors that ease and control discriminative choices of particles by one type of molecule or another. These factors are mainly determined by the original combination pattern (configuration) of the absorbing molecule. They have become decisive factors in the DNA molecule by evolution. The nutrients here are large molecules, not particles.

In the early living creatures the nutrients are supplied by specific organs and transferred by specific mechanism, which gradually evolved into blood and blood vessels in the advance animal kingdom. The absorbing molecule (or the DNA) is swollen by assimilation of nutrients forcing it to split itself continually and thus carries out reproduction. Internal repulsive forces of like-charges start the splitting enhanced by gravitational attraction with neighboring masses. Attractive Em forces operate absorption. Internal repulsive forces, enhanced by external G forces, operate splitting.

The basic idea is that the forces holding particles together into compounds are electromagnetic, not gravitational. On the other hand, remote materials hold together gravitationally at a distance. Accordingly, reproduction is an attractive electromagnetic process, not gravitational. Decomposition and decay of material compounds, particularly organic molecules, are internal repulsive electromagnetic processes, not a weakening of their attractive gravitational forces. See Ch. 7 again. These are the original duplication rules in cells of primitive life. They are replaced by complex duplication methods of DNA reproduction, which evolved further in the advanced animal kingdom.

Chemical elements and compounds carry out their characteristic selectivity also through the *independence* between their G and Em fields. A chemist can list a hierarchical preferential selection in different reactions for each element and each compound, but not the causality of one preferential choice or the other. It is a matter of *range-and-intensity* competence between **G** and **Em** fields near one competitive molecule or another that discriminates in this type of selectivity. A natural rivalry dominates all over the universal arena. The well equipped and first to arrive has a better chance to win rather than the stronger. Whichever particle achieves a shorter range or an easier access, and has the appropriate qualifications, can join another particle or occupy a vacancy in a molecule. This is a basic rule of evolution. Without satisfying it, no evolution occurs. Occasionally, one particle can push another out of a compound and replace it, mostly with the help of a natural or artificial catalyst. A complete molecule often plays similar roles, both in organic and inorganic domains.

Another important role of the *independence* of G and Em is articulated in communication process among brain cells per Em signals among the cells. Recently it has become possible for man to affect, and even change, the communication course among brain cells by changing the Em field intensity in one location of the brain or another. Intensive communication between brain cells can be prompted by direct radiation or by medication, both of which are electromagnetic. Proper feeding plays a similar role slowly and peacefully, which is the healthiest.

A device controlling either or both the Em and G fields can eventually lead to the control of selective growth in organic molecules, and thus control life evolution on earth. Darwin's theory of evolution considers the climate as a most effective factor. Particularly a long term environmental condition and seasonal steadiness of certain weather items (such as temperature-range, humidity, rain, sunlight, latitudinal pressure, geographic locations, etc.) affect evolution profoundly. The type of available food is most influential. Different conditions have different complex impacts on affecting the G and Em fields and therefore affect evolution of living creatures.

Chapter Ten

Field of particles vs. field of forces

Fundamental Field of Forces, **FFF**, is a characteristic field generated individually by each Elixir.[8] Without interaction however, with a field of another Elixir of its own type, an **FFF** remains dormant and shows no activity at all. Then, it is only **potential field of forces**. In other words, without attraction or repulsion between Elixirs, no forces of any type would ever exist. The **FFF** is operated directly by its producer with no mediator(s). It is inseparable forever from its producer. The flux density of the **FFF** is proportional to the accumulative amount of its producer at any point in space. However, the greater the separating distance between any two interacting-Elixirs, the weaker is the flux density of their combined field, according to the **LFF** (Eq. 5.1). Fields of forces of the same type are **unifiable**. The gravitational fields of individual masses are combined into a **common G field,** while the electromagnetic fields of individual charges produce a **resultant Em field** by their combination.

We must not confuse a field of particles with the combined field of their forces. A field of the first type consists of **material entities**, each of which possesses a mass and a net charge. A field of the second type consists of **forces** generated by these entities. Field of forces assume activity through *attraction* or *repulsion* only with fields of similar Elixirs. It extends its influence, which is in fact the influence of its producer(s), throughout the space within its influential range. Each of the material entities generating the G field is found at the center of its individual field before and after combining with the G fields of adjacent particles into a common G field. Consider a moving material field without changing the mutual locations of its components (as in a rigid compound.) It drags with it the individual G fields of the components and their common G field without changing the mutual locations of their individual fields. The exception is as they undergo interactions with other particles/systems. Meanwhile the individual Em fields (of the same components) and their resultant Em fields undergo a similar procedure as that of the G fields. Exception is as an interaction takes place with another particle/system that may cause neutralization among some (or all) complementary charges involved in the process.

Two important facts must be kept in mind. First, as mentioned in Chapter 7, the Em field of a particle has such a high intensity at its generation that it overshadows entirely the G field of the particle, but it declines *sharply* with distance and rapidly fades away within a few inches, showing *no esse nor ens*. On the other hand, the G field is much weaker than the Em field at first but declines *very slowly* with distance and amazingly its influence persists for astronomical distances despite its asymptotic declination with distance (Fig. 7.2). Stephen W. Hawking has estimated the Em-forces between two electrons as about 10^{42} times greater than the G forces.[9]

A neutralized pair (of male and female charges) stores all their combined Em forces, allowing their connate masses to accumulate into larger and larger masses. The common G field of these masses is not only *intensified* by their *accumulation*, but also *accentuated* by the *disappearance* of the external Em activity of their connate charges through neutralization. This gives an illusive impression as if a greater amount of mass was added into the construction of the new unit.

Second, an important difference is recognized between a field of material particles, on one hand, and their combined field of forces on the other. This difference exists in G-fields and Em-fields alike. The individual field of forces of each particle extends continually into space becoming weaker and weaker with distance while each particle producing these fields occupies only a limited spatial volume, as small as its own volume. The particles may either place themselves apart from each other or cluster in large units, not always homogeneously, within their common field of forces. They can eventually form into a concentrated mass of a large particle up to a core of a massive entity, as with a nucleus of an atom, and the sun in the solar system.

This difference between a field of forces and the field of its producers has an important impact on the interaction processes between particles, and on their mutual locations before and after a combination takes place among their fields of forces. It also plays an important role in the Avkoanic level in space. As depicted in Ch. 3, the Negakoans (negatively charged AVKOANS) in space are distributed equidistant from each neighboring Negakoan in the cosmos. Predictably, they do so in a pattern similar to the atoms of sodium chloride (NaCl) crystal.

A decorative cast of electric bulbs in a garden provides a visual demonstration of a combined field of forces containing the material field of it producers. The bulbs are usually installed discretely with a certain chosen spacing pattern among them. They can never be arranged continually the way their lights seem to extend, even if we put them in contact with one another. They always preserve their individual identity, individual *privacy*. These bulbs represent a field of material particles whereas the individual and the combined cast of their light represent the individual and combined fields of forces related to particles. A still better example is the Odor Analogy given in Ch. 8. The plants in this analogy represent a field of material particles whereas each of their two distinct odors represents one or the other of their two respective fields of forces, G and Em.

Chapter Eleven

Sources of singularities in the field equations

During the last ten years of his life, Einstein was beset by the mathematical singularities manifested in his field equation. All attempts, besides his, to unify the G field with the Em field mathematically and solve the singularity problem have been unsuccessful. These attempts had been doomed *a priori* because of disregarding the unique behavior concerning *unstable* particles. This uniqueness arose from the following four factors:

1. The Em field is not a perfectly continuous field but a **resultant field** composed of many Em fields generated by many individual charges of material entities/particles each of which possesses a mass and a net charge.
2. The G field is also composed of many G fields generated by the individual masses of these entities themselves.
3. Each of these entities generates its own Em field and G field individually and simultaneously, before these fields combine with the corresponding fields of adjacent particles/entities into a **common G field** and a **resultant Em field**.
4. As described in Ch. 6, unification between the Em field and the G field is impossible.

A significant attempt to unify both fields and solve the singularity problem has been done by H. Weyl. However, he declared later that he had to forsake any hope *"that the problem of matter is to be solved by a mere field theory."* [10]

Generally, the Em field is not smoothly continuous and its intensity does not decrease uniformly (whether slowly or rapidly) throughout the space continuum. It is strewn, not always homogeneously, with infinite "singularities" which are in fact the very source of the individual Em fields before combining into a **resultant Em field**. Similarly, the G field is not smoothly continuous, but strewn with infinite "singularities" before combining into a **common G field**. These two fields interlace themselves all over the array of their producers, the locations of which turn out to be the points of singularity in their field-equations.

Due to its long range, the G field is easily mistaken as smoothly continuous. On the other hand, repeated reinforcement of the Em field through the fields of like charges (either all male or all female) has created an illusive smoothness in the continuity of their resultant Em field. Nevertheless, the short range of the individual Em fields and the singularities in their field equation disclose the falseness of this smoothness. These singularities in the Em field are caused not only by the masses but also and mainly by the charges themselves, the very creators of their individual Em fields before and after combining into a resultant Em field. Nonetheless, their connate masses distort the expected smoothness of the resultant Em field. This distortion assumes a form of "ripples" in the resultant Em field similar to those formed by obstacles in the wave motion of water. Recent attempts to unify *weak* and *strong* interactions with the Em field through a new theory known as the Gauge theory have been crowned with exciting success. Unfortunately, this unification did not, and could not, include the G field. Moreover, it does not pertain to distinct types of forces. The Avkoan Theory revealed that *weak* interaction and *strong* interaction are **special ranges** in the spectrum of the Em forces. So the said unification has "unified" two ranges in the (same) spectrum of fundamental forces with a third force to which the first two themselves belong and of which they are already parts. These three forces are not different types of fundamental forces.[11]

Quite harsh was the route, and still is, to admit that weak and strong interactions are integral parts of the Em forces. Perhaps unification step is a historical necessity to ease admittance of their inclusion within the same spectrum.

Chapter Twelve

Limitations of our mathematic systems

Formula 5.1 represents quite well the Law of Fundamental Forces. This formula is extendable to harmonize with the requirements of the field theory and the theory of relativity, but this will not be done in this treatise. None of our present Math systems can deal properly with *unstable* particles, not only with quarks and AVKOANS, but also with *all unstable* particles. An extension of formula 5.1 is therefore redundant as far as our concern is to trace matter back to its ultimate particle, the AVKOAN. Precise measurement of mass or charge is necessary. Quantum mechanics can provide only approximations but not precision. Unstable particles always burst out of every violent interaction. Their eruption undergoes a chaotic form and causes the so-called *broken symmetry* or *hidden symmetry*. *Prima facie*, two procedures termed "renormalization" and "gauge field theory" presumably succeeded in eliminating the broken symmetry and solved the problem on an abstract mathematical basis. However, these procedures could not solve the problem on the material (quantum) particle level. No data, no reliable information and not any knowledge could be extracted about the state, behavior and the huge amount (number) of unstable particles that erupt in a violent interaction.

Our remarkable achievements in Math seem to overcome most mathematical problems. If we list all the various branches that we have already covered in Math, we would be amazed how far our knowledge has come. However, now and then, we are confronted with some complicated problems (like the singularities in the field equation) which remained unsolvable by our present knowledge in Math. At least two crucial problems invalidate the applicability of our present knowledge in Math:

1. We are dealing with particles of extremely small masses. Their gravitational activity is therefore negligible. They would virtually surrender to neighboring massive particles that afflict immense forces upon them. Their preference to which of many massive particles they would surrender is impossible to predict. Yet turmoil of attractive and repulsive forces around them stultifies any of their attempts to do so. The net charges of erupted particles are so intensive that turmoil of forces dominates every interaction. The relative distances among the particles undergo continuous change and their effective forces follow suit and change accordingly. Besides, all unstable particles undergo a sequence of interactions in a very short time and change their identities as often. Our available Math systems therefore, including quantum mechanics and gauge theories, fail to help us in dealing with unstable particles.

2. Our present knowledge in Math is far from being satisfactory. I cannot forget an amazing story I heard from the Math

instructor in the first year of my study at the Hebrew University, Jerusalem. He told our class about a famous mathematician who gave a remarkable lecture at a conference. "The audiences" our instructor said, "were so excited that at the end of the lecture they applauded for a long time before the lecturer raised his hands for silence and said: *'Another thing I have to tell: What we know in Math is nothing compared with what is still to be learnt.'* The audiences laughed and applauded loudly again. He raised his hands again and said: *'I heard someone said that I meant an epsilon (ε). This is incorrect; by nothing I meant zero, not an epsilon! That is all what we know in Math.'"*

We were all stunned but none of us dared ask any questions. I confess that I could not accept this last remark of the lecturer until I was confronted with serious problems of particle behavior, which impeded the publication of my book entitled AVKOAN THEORY. Now I can imagine what that eminent mathematician meant. I was aware how far we could handle particle behavior mathematically and therefore I preferred to postpone the mathematic representation of The Avkoan Theory. Nevertheless, I did not forsake trying to put this theory in mathematical terms. Yet I would rather solicit some colleagues to take part in it as well, in the hope that one-day, someone may be able to discover/invent the right Math system that would handle particle behavior properly. Anyway, we must not allow this deficiency to impede natural development of scientific knowledge.

We must keep aiming at our goal to establish what is termed the **Master Theory** despite impediments imposed on scientific research by insistent demanding mathematical representation of particles' physical behavior. Meanwhile, guided by theoretical achievements, the search for an appropriate Math system will continue toward finding one that will serve our purposes properly.

One objective of The AVKOAN THEORY is the discovery of the ultimate particle, the AVKOAN. Precision and accurate values of mass and charge are necessary. For this purpose we need precise measuring devices of mass and charge or either of them in the Avkoanic scale. Devices such as these are still unavailable. In this capacity, extension of Formula 5.1 is useless and therefore redundant. We must not forget that particles are not "tame" entities. They are not indifferent possessions of our Math systems. Each particle has a dynamic mass and a dynamic charge. They assume sturdy fundamental laws of their own. They do not comply with our laws, but ours must comply with theirs. Thus, our Math systems rarely meet their laws. Perhaps, we could have managed to establish some natural laws coincidental with theirs.

In his book *The Cosmic Code*, Heinz R. Pagels provided a profound debate about quantum logic and quantum reality.[12] The reader will benefit very much from reading Chapters 8 - 11 in it. One can judge for oneself then how far we may apply our present Math systems in describing particle behavior.

Part Two

NEW PERCEPTIONS IN THE MICROWORLD

Yecheskiel Zamir

Part II, **New Perceptions in the microworld,** *comprises only two chapters (13 & 14). The first, Ch.13, introduces a deeper insight of heat and temperature. The second, Ch. 14, makes a comparison between force and energy, revealing improper perception of the notion energy. Alternative definitions are suggested to harmonize with* **the principle of the tiniest particle** *and eliminate confusion. The closing Appendix A serves a handy reference to this principle, sparing the necessity of consulting the original. A short paragraph on thermometers is given in Section 13.8 to memorize their historic development before introducing section 13.9. Readers who are acquainted with the history of thermometers may skip Section 13.8 without losing the intended message in Section 13.9.*

Yecheskiel Zamir

Chapter Thirteen

On heat and temperature

> An immersed body in a liquid is lifted upward by a force equal to the weight of its displaced liquid.
>
> Archimedes principle

13.1. Do we really know what is heat and what is temperature?

Before I answer these questions, I must remind the reader of the fantastic temperature currently believed to be dominant in the early universe at the time of the big bang. Scientists have estimated it to be around a billion degrees Kelvin.[13] No one has encountered such a high temperature in the present universe. However, scientists deduced it from the microwave background radiation of the big bang. This radiation was discovered in 1965 by Penzias and Wilson and is found to have a wavelength corresponding to 3° K. The temperature of the early universe was deduced by *extrapolation* back in time in a graph showing the temperature of the universe as a function of time and appeared to ascend to a billion degrees Kelvin. Inevitably, new questions arise: How far can this *extrapolation* method stay reliable? How do we know that no break down occurs before the temperature reaches, if ever, just a few thousand degrees Kelvin? More questions related to *force* and *energy* will be posed in Chapter 14. They will increasingly disprove the authenticity of this *extrapolation*. Perhaps a stumbling block reflects back to **imperfect definitions** of heat and temperature. This Chapter will also consider material involvement in the origination and flow of heat, and whether or not different origination of heat has different impact on matter.

> "Eureka," shouted Archimedes repeatedly as he broke out of his bath around 250 BC, declaring that he found the truth whether or not Heron's crown was made of pure gold just by weighing it plunged in water.

Had the brilliant Greek Scientist found the true nature of **heat**, I believe our modern scientific knowledge would have been quite different.

Heat is one of the most abundant phenomena in nature, and the most influential on life, plants and inert objects. It can be generated from various origins such as sunlight, cosmic rays, light rays, fire, chemical reaction, friction, electric current, electromagnetic waves, etc. Generators that involve nuclear interactions have a unique category, one in which the **temperature conception** becomes **meaningless**. Being a measure of *heat-level*, **temperature** has become even more important than heat itself, because it appears to govern most changes in nature. So, *temperature* is considered as one of the most fundamental notions in science. Its rank in the hierarchy of fundamental notions has probably surpassed that of *energy*. Here lurks the basic problem of why a correct definition of **heat** has lingered behind other scientific ideas. Still, its current definition is inadequate to cover its real essence, its true entity. The (false) belief, that **heat and temperature** (H&T) are most fundamental ideas, has deterred scientists from looking for a more fundamental basis to define them. The definitions of H&T appeared so perfect that no one thought of necessity to amend them. These ideas have been neglected just because of considering them to have reached a prominent rank in science. So no one would dare to discredit them or belittle one's obligation to venerate what they are believed to be.

13.2.1. What is heat?

Only in the mid 19th century the ancient *caloric theory* was abandoned. Up to the 1840s, **heat** was considered an invisible weightless liquid called *caloric* transmitted from one body to another by conduction. Now it is believed to be a kind of energy called *heat energy*, equivalent to mechanical energy according to the Joule experiment. To express the feeling of hot, warm and cold various **thermometers** have been devised to measure the **heat-level** (temperature) of a system and satisfy scientific requirements. Normally, *heat* does not stop transferring from one system to another even when equilibrium in *heat-level* seems as attained. At thermal equilibrium, *heat* exhibits mutual exchange in an isolated system by conduction and/or convection of small particles and molecules. *We* will see that heat is a **collision** of particles that escape, or are expelled, from the atoms in a chemical reaction.

13.2.2. Apparent mystery: The mystery behind **heat conception** has never been revealed. All we know about heat is only a description of its apparent impact on human senses when a material entity is exposed to heat, just as any living creature would *react/feel* when it encounters heat. We do so only in the abstract perception of heat. Yet, no one tried to consider involvement of real material-entities uncurbed by imposition of our imaginative abstraction. Today, **heat** is still recognized according to its *external* effect on material entities, ignoring its impact on the *internal* structure of matter during a heating process. The true nature of heat must be explored in its successive reflections on the *innermost* constituents of matter. To wit, we must direct our exploration into the **inner structure** of *atoms* and *molecules*, concerning heat. Heat shows no different impact due to its origin. However, the heat originated by nuclear interactions enters a stage beyond a **critical point** that will be discussed in Sec. 13.9.

13.3. The heating drama

When a combustible material decomposes, the electrons revolving around its atomic nuclei get *excited*. They set free from their orbital motion by their own repulsive forces, bounce violently from the atoms, and normally burst into fire. Let us identify these electrons as **initial electrons**. They collide with nearby substances and induce *excitation* in the electrons revolving around their nuclei as well. These new electrons are to be identified as **secondary electrons.** Their

excitation is recognized as **warming.** Little difference could be between natural and artificial heat regarding their impact on nearby material. The greatest supply of natural heat comes from sunlight, providing *initial electrons*. Next to it stands radiation in different frequencies. Only in the third degree comes volcanic fire and combustible material that burst into fire without human intervention. Artificial heat involves human intervention, whether by friction as was raised by man in the stone age to produce fire, or by a modern supply of fuel. Once fire starts, the heating process proceeds in a similar way whatever its origin. It affects matter immediately after *excitation* of secondary electrons when they are still in their orbits around the atomic nuclei, i.e., before the electrons begin to escape from the atoms. This **excitation** is a warming condition. It brings the system into three dramatic **heating steps.**

1. **Atomic swell:** After *excitation* by initial electrons, the *secondary electrons* begin to repel one another within the atoms. They are compelled to accelerate and enlarge their orbital paths around the nuclei. The atoms *swell* accordingly and impart expansion to the entire body in a warming condition. (The coefficient of expansion differs for different elements and depends on the structure of the atoms and on the mutual orientations of orbital motions of the electrons in their shells.) In this condition, just a little more heat is enough to cause detachment and escape of *secondary electrons* from the atoms.

2. **Exodus:** When *secondary electrons* escape from the atoms, their mutual repulsion becomes intensive. They collide at the surface of the mother-material and force other *secondary electrons* to release themselves from the atomic structure. Discharge of electrons from a body/substance induces a "feeling" of loss of these parts. How? Losing parts creates vacant spaces inside body/substance and triggers turmoil of displacements of electrons, atoms and molecules. The entire heated body undergoes successive changes. Some changes are mild. Others are drastic. It turns out that **heat** is *a loss in the atomic structure of a body* under violent collision inflicted by boisterous free electrons that cause internal changes. A higher **rate** of this *loss* is recognized as a higher **temperature.**

3. **Atomic decomposition, bare atoms and bare molecules:** Several atoms become **bare**[14] of their protecting electron-shells, i.e., they become *positively* charged so that repulsive forces are created among them as well. They detach themselves from the mother-material and collide with anything in their way. The more heat is applied to a system, the more *secondary electrons* get loose and escape from the atoms leaving the atoms **bare** behind them. An additional heat to the system would increase the number of free electrons and boost their repulsive forces. The *bare atoms* undergo similar repulsion fate. A higher *density* of repelling particles is

recognized as an intensified **heating process,** which involves additional decomposition of the atoms into electrons and bare atoms. It destroys the chemical bonds between the molecules. "Chunks" of *bare* (positively charged) molecules, beside *bare* atoms, begin to detach from the heated substance. Mutual repulsive forces of their own also accelerate these molecules. Ever more molecules are set free by their repulsive forces until the entire heated material changes phase, solid into liquid and liquid into gas. As heat continues, electrons of the **inner shells** begin to emerge from the atoms. Now the atoms get into a higher stratum of decomposition. Electron detachments from *inner shells* impart roaring mutual collisions until the entire substance evaporates into *bare atoms and bare gas molecules.* They escape from the system, where they get neutralized with free electrons of earlier escape.

13.4. Normal heating

At normal heat, matter decomposes into three groups of particles:

1. *Free* negatively charged electrons.
2. *Bare* positively charged atoms (unprotected by outer shell electrons.)
3. *Bare* positively charged molecules, whether in a liquid or gaseous phase.

This procedure is relevant only at **macroscopic** level that excludes nuclear interactions. Normal heat is limited to chemical reactions and transformation of one physical phase (solid, liquid and gas) into another. When heat involves nuclear interactions, it enters a **microworld**, the world of *subnuclear* particles that is out of our discussion in this limited topic of heat. One or the other of the following behaviors is determined by the atomic structure of a heated material:

1. Combustible behavior: Enormous **initial electrons** get excited and simultaneously escape from the atoms in a vigorous discharge. The heated substance burst spontaneously into fire and flame, biasing or leaping over a warming step.
2. Normal behavior: The state in which three *heating-steps* take place, as described above in the Heating Drama.
3. Nuclear reaction/radiation: This behavior is characteristic of radioactive substances. It can also be prompted by X-rays or laser beams, which penetrate the atoms, hit their nuclei and split them up with or without explosion.

In short, **warming** begins with *excitation* of electrons, within the *atomic structure*, and ends in their calmness while they are still rotating around the atomic nuclei. **Heating** begins with electron detachments from the atoms and ends in neutralization of bare atoms and bare molecules. The subnuclear stage excludes the temperature idea entirely. When the bricks of a building get loose by some force, whether internal or external force, they fall down to the ground by gravitational attractive force. The velocity and the number of bricks that fall per second determine the *destructive power* they can cause by their collision with anything. In the heating process, on the other hand, the more heat applied to a system, the more electrons get loose and escape from the atoms. Continuous heat releases ever more electrons and boosts their repulsive forces. So, they get ever more speed and their *destructive power* at collision increases accordingly. Heat and temperature are strongly related to a fundamental idea termed **force**. The *fundamentals* of **force** precede those of heat, temperature, and even energy. None of the latter three ideas could exist without force. Only a force can impart palpable material involvement and concrete meaning to these ideas.

13.5. The cooling drama

When a heated substance decomposes it undergoes immediate chemical reaction with adjacent compounds. Any new combinations would soon decompose unless the heating process stops. *Bare gas atoms and bare gas molecules* escape from the system, attract free nearby electrons, get neutralized with them, restore their atomic and molecular structures and calm down. **Neutralization** is a **cooling process**, because it imposes **slowness** in mutual collision. It occurs around the system even during a continuous heat. When many electrons escape from the system the density of the remaining free electrons decreases and their mutual collision slows accordingly. *Collision-slowness* is an inevitable step of a *cooling process*. It is recognizable when the reduction in the density of free electrons is not compensated by additional supply of heat. *Collision-slowness* imparts a sense of **gradual cooling** of the whole system. Cooling undergoes a specific drama. The electrons in the atomic shells of a cold body slow their rotation around the nuclei. Their orbits shrink and the entire body follows suit. Persistent cold develops **freezing,** which is normally a solidifying process. Severe cold enfeebles the human body. The freezing process is simply an *extreme slowness* of electrons in their *orbital motions* that could drop down to a very low frequency, but they never become motionless. Such a reduction, or almost lack, in motion could end in the death of living creatures. Normal cooling follows immediately after heating stops, because bare atoms and bare molecules begin neutralization with free electrons.

13.6. Abstract character of heat and temperature

Heat and temperature are abstract ideas. They are as abstract as the ideas distance, height and volume, to each of which a convenient **zero-point** and a **unit** (and a *direction*, too, where necessary) are assigned to measure their quantities. Any fixed point on earth or in space, for instance, can be taken as the *zero-point* along with a well-defined *unit of length*, for measuring distances. For height, the sea level has been (or the bottom of a container is) chosen as the zero height-level along with a convenient unit of length. For temperature, the freezing point of distilled water (at normal pressure) has been established as the zero-point in the Centigrade Scale, while its boiling point has been satisfactory to establish the unit (degree) of temperature. It is only a matter of convenience that water has been chosen to serve this purpose. Any other (material) substance can serve the same purpose. Abstract ideas and arbitrary units have also been adopted for measuring time, speed, density, weight, angles, heat and energy. The units of these ideas are more complex than those in the preceding group. Some abstract ideas like "chair, room and color" are intended to signify functions or feelings but nothing about the involvement of (the type of) a material entity. A (material) stone, for instance, can serve as a chair, a table or a hammer; a cave can serve as a room; coloring powder or (material) vibration as color; and so on.

As practical a service and as precise a measurement as these ideas and units may provide, they can never replace the material stuff involved, but they do promote the implications and forms of matter in our imagination. The idea distance, weight, temperature, heat, energy or the like is *not real matter* itself, although each of these is related to material objects. Likewise, their units, such as a mile, a pound, a degree, a calorie and/or erg, are *not real* objects.

This discussion is very important, not only because of the abstraction of **heat and temperature** (H&T) but in particular because of the combined abstraction of **energy and temperature** (E&T). No one has dared to degrade E&T, and no one would, regarding their roles in describing material behavior, as far as no shortcoming on their part has been found that may refute their significant roles. Their prominent scientific status requires full compliance with them by any phenomena and any law that involve E&T to earn a firm establishment in the scientific community. Energy and temperature have been so excelled in the hierarchy of fundamental ideas that they became *the* primary tools in describing the creation and evolution of matter and of the entire universe. Yet, according to **the principle of the tiniest particle** (see Appendix A), **mass** and **charge** are not only *the* most fundamental qualities in nature, but they are also the *only* two attributes of matter. *Heat, temperature* and *energy* are abstract because they are immaterial qualities. Unless they are defined by the material particles involved, they stay abstract although they describe the state (in which matter may be) and/or other properties about matter.

13.7.1. What is temperature?

Macroscopically, **temperature** is the **heat-level** of a system. Still, this definition is incomplete and requires replacement or amendment that expresses the **excitation rate** and **detachment rate** of electrons. An electron detachment however, is inevitably followed by detachment of *bare* atoms and *bare* molecules (Sec.13.3, Step 3). The more free electrons per unit volume, the more collision they exhibit and the greater speed they acquire. This **speed** determines the *temperature,* or the *temper*, of the system. Repulsive electromagnetic forces manifested as chemical reactions normally cause a detachment of particles. Microscopically, *temperature* becomes **meaningless**, because of involvement of free *subnuclear particles*. The denser these particles are, the greater become their repulsive forces. Also, the more massive they are the greater impulse they imply. This impulse is powerful enough to set them free and let them escape out of the system, despite their *heavy* masses, compared with the electron mass. They attack any material object that stands in their way. They interact with it and decompose it. Mostly, nuclear interactions involve explosions, expelling enormous subnuclear particles out of a system. Just as with electrons, *dilution* of whatever free charged particles brings a system into a cooling procedure. ***Temperature*** *expresses behavior only of **lepton** charged particles, particularly electrons.* Bare atoms and bare molecules are indirectly influenced by temperature through ***lepton*** particles' activity. The quantity of heat in a system resembles the quantity of liquid in a container, and the *heat-level* (temperature) resembles the height of liquid in a container. An amount of liquid in a narrow container reaches a higher level than it does in a wide one. Analogically, a quantity of heat in a small body rises to a higher temperature than it does in a massive body.

13.7.2. Indeterminate Domain:

Temperature becomes indeterminate in subnuclear domain and so is the notion heat. Accordingly, they become **meaningless**. One can easily recognize that, in essence, *the prevailing perception of heat does not differ much from the ancient caloric theory.* The introduction of temperature in the 18th century could fit quite well as caloric-level in the ancient heat-perception. Unfortunately, the discoveries of subnuclear particles attenuate the comfort offered by the conceptual introduction of temperature. The current view of temperature is *unsuitable* in the subnuclear domain, unless we replace or properly amend its definition. I must emphasize again that the whole idea of temperature becomes **meaningless** in a subnuclear domain.

13.8. Thermometers

In 1592 Galileo invented the first thermometer. This was a gas thermometer that lacked precision. Several improvements followed in the 17th and the 18th Centuries until the mercury thermometer was designed in 1743 with Celsius Scale, which today is a standard scientific thermometer. The *freezing and boiling points of distilled water* at normal pressure were chosen as **fixed points** for gradating thermometers. In the Celsius Scale, the freezing point of water was assigned as $0°C$ and its boiling point as 100°C. This interval between FP and BP of water was divided into a hundred equal sub-intervals each of which indicated one-degree Centigrade. In the Fahrenheit Scale, the same interval between these two points was divided into 180 sub-intervals, so that $0°C$ would correspond to $32°F$ while $100°C$ would correspond to $212°F$. Distilled water has genuine freezing and boiling points at normal pressure. Fortunately, nature has bestowed upon us water abundantly on earth, in rivers, in oceans and as vapor in air. A huge amount of rainwater penetrates the soil and is stored beneath the surface of the earth serving as springs and reservoirs. It may be drawn up from wells or cisterns. So, thermometers can be constructed and gradated easily everywhere on earth. The water must be distilled ahead to fix up its freezing and boiling points on thermometers for gradations. Various types of thermometers are in use today. Some of them are designed for temperatures below the freezing point of mercury, such as alcohol thermometers. Others are designed for temperatures above the boiling point of mercury, such as a thermocouple. Nevertheless, **none of these thermometers is perfect.** Each of them has advantages and disadvantages of its own. Specifically, the mercury thermometer begins to fail in time and requires a new gradation every three to four years. The choice of measuring material depends on the temperature-extent a thermometer is intended to serve, whether low, medium, high, very low, or very high. The following properties of measuring material are crucial to design thermometers:

1. Freezing and boiling points.
2. Coefficient of expansion.
3. Conductivity (to set sensitivity).
4. Phase, (whether it is a gas, liquid or solid, at normal pressure and room temperature.) For liquid, the container coefficient of expansion must match that of the liquid, as in mercury and glass.

13.9. A critical point

Above a certain critical temperature, matter decomposes totally into subnuclear particles that burst out of the system with extremely high speed and collide violently and chaotically with one another. They would respond no more to *temperature*. Beyond this critical point, the *temperature conception* loses its authenticity. Then, we can speak only of **mass** and **charge** that, according to the **principle of the tiniest particle,** are the only two attributes of matter (see Appendix A). A state as this resembles that of the early universe when only primordial and subnuclear particles predominated. **Heat** and **temperature** apply only for material entities in the atomic stage and above it. Once the nucleus decomposes, heat and temperature are no longer relevant. New ideas are required to replace them for describing particle behavior in the *subnuclear stage*. Heat and temperature are only **consequences** of *composing and decomposing* matter. The basic idea is that **H & T** do not govern material behavior as currently believed. Neither **energy** does. On the contrary, these three ideas are governed by (material) particle behavior, which in turn is governed by fundamental (attractive and repulsive) forces among charges. Confirmation of this statement requires exploring particle behavior in a state of excitation and high concentration of subnuclear particles that produce a current of these particles. Such a state is achievable under high frequency radiation or under bombardment by high-speed particles as are produced by high-speed accelerators. High-speed particles imply high frequency rays, all of which presumably propagate at the same speed of light. Heat involves **charged particles in motion** resulting from forces created by their characteristic attraction and repulsion. Their speed is determined by the intensity of these forces according to the Law of Fundamental Forces, **LFF**, as shown in Formula 5.1. The type of material reaction with a particle is determined by the particle structure. However, particles behave differently in different compounds. The release of electrons from an atom depends on the intensity of their "confinement-force" in a compound. On the other hand, free electrons respond immediately to heat. Relating *heat and temperature* to particle behavior is a logical step arising from the normal heating steps described earlier. Two advantages stand behind that:

1. It fortifies renunciation of extending the current definitions of H&T over all stages. Extension has failed in the subnuclear stage.
2. It eliminates involvement of temperature in the early universe.

Heat and temperature are improper tools for describing particle behavior in the subnuclear stage. Their definitions require replacement by new ideas, rather than amendment, to meet heat responses at different stages, independently of the notion heat or temperature, as follows:

1. **Warming stage:** A stage in which the orbital electrons around nuclei get *excited* without detachment from the atoms.
2. **Soft chemical-stage:** A stage of gentle chemical reactions and moderate electron *detachments* from the atoms, without fire.
3. **Burning-stage:** A stage of strong chemical reactions accompanied violent escape of electrons (and other material fragments) in huge quantities exhibiting fire and flame.
4. **Subnuclear-stage:** This is the stage of nuclear interactions. Here, the ideas of heat and temperature become **meaningless**.

In a WARMING STAGE, a heated substance preserves its molecular bonds and its atoms preserve their (atomic) structures. No chemical reaction takes place in this stage. Heat may be defined here as an **enlargement** of electron orbital paths in the outer atomic shells, temperature as the rate or the **frequency** of these orbits.

The SECOND STAGE involves gentle, slow chemical reactions with slow occasional **detachments** of electrons from the atoms but without fire. Here, heat and temperature concur with their current definitions, preferably to mention involvement of electron **excitation** within the atoms. The formula $Q = mc(t_2 - t_1)$ of heat quantity is valid.

The BURNING STAGE is characterized by strong chemical reactions and spontaneous *escape* of electrons from their orbital motion around the nuclei, which may be controlled, stored and reproduced at will. It creates a current of free electrons, as by batteries, electric generators, lightning and carpet-static. Their own mutually repulsive forces monitor escape of electrons. In this stage, the chemical reaction is accompanied by fire resulting from electron currents or oxidation, particularly with oxygen. Here, a definition **amendment** is necessary: *heat* may be defined as an **escape** of electrons from their orbital motion around the nuclei, *temperature* as the number of electrons that **escape** from the atoms *per unit time-interval* or as **frequency/vibration** of free electrons. Nevertheless, the heat quantity formula $Q = mc(t_2 - t_1)$ still valid.

The SUBNUCLEAR STAGE implies a microworld in which nuclear interactions take place so that the ideas of heat and temperatures become **meaningless**. Heat and temperature have no control on subnuclear particles. Therefore, they are no more authentic but require **replacement**, I propose, by new ideas that express the number of *free subnuclear particles per unit volume (density)*, dispensing with using improper notions as heat and temperature. This will not only improve description of particle behavior at the early universe, but also save us embarrassment of absurdity in surmising *illogical* fantastic temperatures to the early universe.

13.10. Matter vs. temperature

As described in Ch. 3, attractive and repulsive forces among charged particles play an important role in the creation and evolution of matter and the universe. Attraction prevails within a chemical combination. Repulsion prevails outside it. Both activities exist everywhere. Yet, their responses to temperature differ in different generations, as classified ahead:

1. Subnuclear generation, manifested in creating elementary particles out of primordial particles, dubbed **AVKOANS**.
2. Nuclear generation manifested in creating hadron particles.
3. Atomic generation manifested in creating atoms of different elements.
4. Molecular generation manifested in creating molecules up to galaxies.
5. Organic generation manifested in creating botanic elements and living creatures.

Subnuclear generation: This generation begins the creation of elementary particles like quarks and leptons out of primordial particles termed **AVKOANS**. Repelling forces prevail among like charges compelling their carriers to flee from one another. The greater concentrations of like charges per unit volume, the more vigorous they become under the influence of their repulsive forces. Material combinations, including neutral particles, cannot maintain their combined forms in this generation. High-speed particles attack and decompose them immediately before they attain stability. Microworld is a chaotic world where particles cannot be orderly but in their absolute compliance with the Law of Fundamental Forces, **LFF**, (Formula 5.1). The existence of this law arose from *attraction* and *repulsion* values between masses and among charges, without which matter could not be created. Microscopic particles are too wild in their free state but they comply sturdily with **LFF**. However, when neutralized into stable units, they sublime into the macroworld, adapt and conform to its entire orderly rules, verbatim. For then, they no longer belong to the microworld. **Subnuclear particles are uncontrollable by heat and temperature**.

Nuclear generation: Attractive forces prevail between unlike charges inciting their carriers to move toward each other until they unite and calm down at the formation of neutral units. Neutral particles are indifferent to external charges. Yet, they maintain gravitational activity according to the Law of Fundamental Forces, **LFF**. Each pair of balanced charges conserves its *electromagnetic forces* within a new combined unit. These forces conform to some rules that are not clearly understood now. Some scientists believe they are. Anyway, these forces turn out to be the **binding forces** of the new units. Binding forces are erroneously considered binding energy. A concoction of subnuclear particles starts in this generation arising from ample neutralization of charged particles incidentally. They undergo further interactions with one another to form into particles called **hadrons**. Among these are protons and neutrons (nucleons) which are the basic constituents of the atomic nuclei. Chaotic collisions of the preceding generation drop deeply into mild occasional collisions. Still, temperature conception is inapplicable in this generation too.

Atomic generation: This generation starts with *simultaneous* neutralization of enormous particle-pairs, causing a general *slowness* in collisions that prevailed in the first and second generations. This *slowness* soothes the remaining free particles. Nucleons of the preceding generation combine into a variety of nuclei, attract electrons created in the first generation, and form into atomic elements. As mentioned earlier, atoms respond to the heat-phenomenon. From this generation onward the notions of heat and temperature begin to make sense and continue to do so in the molecular through the organic generation.

Molecular generation: Except hydrogen atoms, all atoms require completion of their outer shells into eight electrons to attain stability. They do so either by receiving electrons from other atoms or by giving up their outer shells electrons. In the first case they become negatively charged, in the second case they become positively charged. The number of electrons each atom exchanges determines a multitude-charge of each atom. To recover and maintain neutral states, atoms undergo chemical reactions with other atoms. All the atoms in the new combinations attain individual and collective neutralization while they complete their outer shells into eight electrons (except hydrogen atoms, each of which requires only two electrons in its single shell[15]). These reactions contribute to stability. Some compounds attain strong and steady stability. Others are too weak to stand against external attacks in normal weather conditions. Most atoms share their outer shell electrons covalently with other atoms in the formation of compounds. Ever more molecules are produced. Their quantity increases and eventually they evolve into large material substances up to galaxies. Subsequently they expand into space within billions of years. However, their gravitational activity holds back their expansion. What happens further is still under debate. Heat plays an important role in this generation, and so does temperature.

Organic generation: Thanks to the formation of water molecules in the Molecular Generation and oxygen in the Atomic Generation, rudimental botanic elements managed to form and grow in water, and then on soil, too. It took billions of years for botanic elements to evolve into a variety of vegetation, trees and living creatures. In turn, an additional interval of billions of years was required to evolve into a variety of aquatic elements, microorganisms, and animals up to human beings. Temperature plays a crucial role in this generation because plants and living creatures can survive only in a very limited weather condition and a very narrow temperature interval.[16]

This generation requires specific conditions available on planet Earth. Unless such proper conditions exist elsewhere in the universe, no living creatures could be expected there.

Chapter Fourteen

On force and energy

14.1 Force

What is a force? Is it something that exerts pressure like muscles, or is it something that causes motion? The definition of **Force** has underwent several revisions in the history of science. None of these definitions has been satisfactory. In the mid-twentieth century "force" has been given its current definition. It is anything that causes a change in velocity [a change in magnitude and/or direction of the speed of a body, whether it is at a relative motion ($v \neq 0$) or at rest ($v = 0$)]. Simply, it is whatever *accelerates* or *decelerates* a material entity. However, this definition is still incomplete. It does not satisfy the requirements of fundamental forces. Forces may be classified in two major categories, **fundamental** and **non-fundamental**.

As perceived by The Avkoan Theory, a **Fundamental Force, FF**, pertains to one of two functions, either **attraction** or **repulsion**, among masses or among charges. Each mass generates a gravitational field (**G**), and each charge generates an electromagnetic field (**Em**), of forces. These are the two **Basic Fields** of matter and are the only two Fundamental Fields of Forces, **FFFs**, in the universe. The first is due to the existence of mass; the second is due to the existence of charge. Basic Fields inherit their fundamentals from their producers, **mass and charge**, without which matter could not exist and, obviously, forces as well.

Non-fundamental forces are not immediate interactions among masses or among charges. So all forces other than the two basic fields are *non-fundamental*. They assume various forms, like tensions in muscles, springs or strings, mechanical forces (including mechanical centrifugal and centripetal forces), forces governing winds (including cosmic winds), etc. These forces are *not* immediate interactions among masses or among charges, i.e., they are neither *gravitational* nor *electromagnetic*. Therefore, they are not fundamental. They can only be produced or derived from the two **FFFs**. Being a *force* per unit area, the notion **pressure** has the same fate. On the other hand, a chemical force, fuel, food, atomic and nuclear forces are electromagnetic because they are immediate interactions among charges and are therefore in the category of fundamental forces. Falling-bodies on earth, planetary motion in the solar system, and the like are motivated by gravitational forces, which are also fundamental. Appendix B stipulates situation where **Marvelous Forces** may exist.

14.2 Energy

What is energy? Energy requires either a charge in motion or a mass in motion. It is often confused with force and vice versa because the definition of **potential energy** is distorted. This is not energy at all. It is a force, just as the force of a stretched or compressed spring. It is similar to a force in a human hand (muscles) carrying a heavy weight without moving with it.

The term **potential energy** is related to a force that holds together the components of a particle or a system of material objects. However, the modifying function of the adjective *potential* has faded away in time by improper use, so that the notion of *force* is confused with the notion of *energy*. Only forces in a state of tension hold together the components of a particle/system, *not energy*. Consider a body on a table. An attractive gravitational force exists between the body and the earth, but *not energy*. Due to gravity, the body tends to "fall" toward the center of the earth, and so does the earth toward the center of the body. However, the mass of the earth is too large for this force to get it moved discernibly, so that the body alone makes a discernable motion toward the earth. Holding the body from falling toward the earth, the table stands as an obstacle between the body and the earth and *suffers* stress from the tension prevailing between them. Unless the table cracks under the burden, it will keep *suffering* from this tension as long as the body is on it. When the obstacle is removed or the body is pushed away from the table, the body resumes its motion toward the earth and thus displays energy known as (gravity) potential energy, as given in Formula (14.1). Another formula (14.2) is given for *elastic* potential energy related to a spring. The subscript p denotes here a potentiality.

$$E_p = mg (h_2 - h_1) \quad \text{............(14.1)}$$
$$E_p \text{ (elastic)} = \tfrac{1}{2} kx^2 \quad \text{............(14.2)}$$

However, a potential energy loses its *potentiality* as the body falls/moves. Then it becomes a *factual* energy, which we call **kinetic energy**.

Like every material entity, the body possesses nothing but **mass** and **charge**. On the table, the body is under the influence of gravitational attraction as tension. This tension is a **force**, not energy, prevailing between the body and the earth. It holds both of them together just as a spring might hold two rigid objects. It is not a property of the body alone. Without the earth, the body cannot *store*, or engage in, a gravitational force either. During its fall the body does not lose but looks as if it gains **force** but not energy, because its distance to the center of the earth becomes shorter, according to Formula 5.1. In fact the body neither loses nor gains anything as it falls. It maintains properly the only two things that it has, its *mass* and *net charge*. Nevertheless, to lift it, one must apply a greater force on it to overcome its weight, i.e., the gravitational force exerted on it by the earth. This force (the weight) increases continually as the body falls.

A body in motion displays energy termed **kinetic energy** that considers only mass and velocity. When the body is moved, a force must have caused this motion. The force spends energy termed **work** or **mechanical energy**. This energy is in direct proportion to the body's weight (mg) or the force (ma) exerted on it, and to the distance of displacement. Often people say: "Mr. So-and-so has energy in his body." They should say: "...has a force in his body." Others say: "food, stores energy in a body." They should say: "...stores a force (or a fuel) in a body." These forces are destined to display energy only when they are set in motion. Otherwise, they stay **dormant forces** being stored in the body, but not energy. Food may supply energy just as a fuel does as it burns. To lift something up to a certain height-level, one must spend energy, which is *mechanical energy*, not potential energy. It is erroneously considered *as if* recoverable when the body falls back to the same level from where it was lifted. The falling process is associated with energy because the body *moves* a certain distance, not because it has energy. This confusion arose from the conservation law of energy. Once this law took charge in science, the *potentiality* was ignored and the *potential energy* began to be viewed as actual energy despite its limitation by the modifying word *potential*. If we accept this view, all types of *potential forces*, like chemical bonds, electric forces, atomic and nuclear forces, etc., should be treated similarly as energy. In fact all material entities are held together by *forces*. So, they should be treated similarly, as the *gravity* and the *elastic* potential energy, E_p. Yet, this is wrong although forces are destined to produce energy upon their release. **The gravitational force should not be treated as energy.** Without falling, a body has no energy at all, not an active nor inactive, neither a hidden energy. Material body has only a **mass** and a **net charge**, nothing else. A gravitational force is exerted by the earth on its mass expressed as $W=mg$, which is the body's weight. This force is not energy. Similar argument applies to elastic potential energy. When a body is confined to a place without moving, it has no energy and it will not have energy unless it moves. A crane is a dancing bird. Its *potential dancing* is obviously **implicit** in the name, crane, whether it is engaged in actual dance or not, and without explicit expression. Similarly, a healthy person can jump at his will. His *potential jump* is obviously **implicit** in the notion *person* without mentioning that explicitly as *potential jumper*. Avoiding the use of *potential energy* is crucial because using it implies false perception of a *dormant force*. Due to their small masses, electrons have negligible gravitational activity. So, atoms in a compound are considered as rigid parts despite the moving electrons around their nuclei.

14.3 Equipotentiality

Material entities are said to be at **equipotential level** when they are on the same earthen level. When a body moves in the same latitudinal level on earth (right or left for instance) no change occurs in the value of the gravitational force exerted on it by the earth, despite its displacement a certain distance away. However, this movement necessarily expresses energy, not potential but **kinetic energy** $E_k = \frac{1}{2} mv^2$, i.e., a *motion energy*. A force must have caused this motion and must be taken into consideration. Just a small force is enough to move a body over an equipotential level on earth to overcome friction, as a boat on the surface of water. Likewise, an electron movement over an electric **equipotential surface** expresses **kinetic energy, E_k,** which must be taken into consideration as well, not potential energy. Repulsive forces exist between like charges and attractive forces between unlike charges. These forces are measured in volts. When an electron moves from one potential surface to another, it does so due to **electric energy**, which is expressed by the equation E=eV or by its equivalent *kinetic energy* $E_k = ½ mv^2$. Beside gravitational field, a charged particle is surrounded by its electromagnetic *field* of forces, **Em**, not by electromagnetic *energy* as some people think. A material body is set in motion only by a force. A charged particle is set in motion only by **interaction force** with the field of another charge. This motion must involve energy, **electric energy.**

14.4 Other types of energies

Beside work-energy, kinetic-energy and electric-energy, other types of energies are known, which bear various names denoting the sources of their generations, such as thermal energy, chemical energy, photon energy, electromagnetic energy, and the energy expressed in Einstein's equivalence $E=mc^2$. All types of energies that involve charges are **electromagnetic**. As far as these charges are confined without moving, they are engaged in some **static** *electric forces*. They are destined to evolve into energy only when the charges move, normally due to interaction (either attractively or repulsively) with the **Em** forces of another charged material. Beside physical force, various types of *static* forces are erroneously considered as *energy*. The following examples are common:

* Halos that appear around human bodies, vegetation or any material entity,
* Meditation or thinking force,
* Force of sexual attractiveness,

* Force of hatred or love feeling,
* Intuition force,
* Invisible force acting as a magnet in a human hand stretched unto an object,
* A concentrated gaze on something,
- Fuel, etc.

Most of these forces are produced by **overcrowded** *static charges*, which are erroneously recognized as energies. From the presence of static charges arose a false belief that old forests and old graveyards are occupied by energy. Some people consider them haunted by spirits. Generally, decomposition of organic substances discharges static charges. So, old graveyards and forests are saturated with static charges, which maintain some distances apart from each other because of their repulsive forces. Occasionally, they fluctuate in amazing glitter by breeze or slight wind. Just a little excitement would cause a nearby person to emit additional charges from his body that make him appear surrounded by a halo. **Energy-delirium** has conquered the human mind so deeply that most forces are routinely and recklessly *considered energy*. So it is among writers, scientists, and the media. A recent book entitled *The Celestine Prophecy* recurrently expressed the "modern" misleading perception of *energy*.[17] This endeavor did not stop throughout this book, starting from Ch. 3. It has gone as far as "seeing" **energy of God**, (pp. 153 & 236). Reading this book is pleasing, thrilling, and inspires deep rumination/musing. Sadly, toward its end the author inserted a melody of miracles. So he deprived me of the pleasure in which I was absorbed throughout my reading until the mysterious disappearance of a hero named Wil (p 242): "*His image was becoming haze, distorted. Gradually he disappeared altogether.*" Despite my criticism, I recommend reading this book, not only because it is thrilling and inspiring but in particular because it urges a necessity to stop the current distortion of the notion *energy*. And the sooner, the better. Contrary to the prevailing belief, **energy is not a fundamental attribute of matter**. Energy is *produced* by forces that are themselves *produced* by the only two attributes of matter, **mass** and **charge**, (see Appendix A), and so are heat and temperature. Charges are set in motion, with their *carriers*, by interaction with fields of other charges. This motion is responsible for the creation of **electromagnetic energy**. Precisely, **Energy** is only a *product* created by interaction among masses or among charges. Without mass and charge, matter could nowhere be created and obviously neither could *energy*. This is a basic principle of the **Avkoan Theory.**

14.5 Conclusion

We are not entitled to describe the creation of the universe through energy and temperature. A more convenient description would be through mass and charge, based on the **principle of the tiniest particle** declaring *mass* and *charge* as the only two attributes of matter (see Appendix A below). By replacing the idea *temperature* with **density** *of free charged particles*, we save unpleasant confrontation with bizarre degrees of temperature for the early universe, around a billion degrees Kelvin as described in Sec. 13.1. One may summarize the entire Chapter 14 as follows:

Energy cannot exist without motion. Motion cannot exist without force. Forces cannot exist without fundamental forces. Fundamental forces cannot exist without attraction and repulsion among masses and among charges. Precisely, forces of any type, whether physical (material) or psychological (nonmaterial), cannot exist without fundamental forces.

Appendix A

The principle of the tiniest particle

Indivisibility of the ATOM prevailed around 2450 years until 1913 when Rutherford found it composite (of a nucleus and electrons.) Ten years later, the nucleus was found composed of protons and neutrons (1923). Many discoveries of subnuclear particles followed (1940-75). The quark model was raised with its surprising fractional electron-charge (1963). Presently, the AVKOAN PRINCIPLE has ended splitting of matter *forever*. This principle has lifted a banner declaring THERE CAN NEVER BE EXPECTED A DISCOVERY OF A PARTICLE SMALLER THAN THE AVKOAN.

The **Avkoan Principle** states the following:

> *There must exist a group of primordial particles, termed AVKOANS, which are the tiniest, in a sense that **there can never be expected a discovery of a smaller particle**. Each AVKOAN necessarily possesses two intrinsic attributes, which are independent, indivisible, inseparable and indestructible: one **Avkoanic-mass** and one **Avkoanic-charge**. These are the **ultimate units** of mass and charge. Any particle whose mass or charge is bigger than these ultimate units is divisible and*

cannot be tiniest; viz., it cannot be an AVKOAN. AVKOANS must therefore be identical in their masses and equivalent in the absolute value of their charges, each of which can be either positive or negative. Ergo, the entire universe is composed only of AVKOANS.

Appendix B

(Supplementary to Section 14.1)

A Marvelous Force

The *independence* between mass and charge activities brought about the independence between the two Basic Fields **G** and **Em** of every material entity. This independence implies that unification between these fields is impossible, despite their coexistence and their simultaneous generation from the same individual particle in every material entity. They never interfere with each other's activities. The fact that they always originate together and "defuse" from the same originating entity creates an illusion that they are **unifiable**. In actuality, these two fields, **G** and **Em**, are absolutely independent. The only relationship between them is their coexistence and their simultaneity, arising from the inseparability between their originators, mass and charge, which are the only two attributes of matter.

The eternal association of mass and charge in the tiniest particle, and the *simultaneous* origination of both gravitational **G** and electromagnetic **Em** fields from every material particle/system, stimulate a splendid but queer inspiration. It is a *wish*, that the tiniest particle of matter were to consist of only one quality, a mass, which is characterized with both functions, gravitational and electromagnetic; as if this quality (the mass) itself **flashes** an **Em** field additional to its **G** field. But this fancy is inconsistent because the mass and charge of every material entity assume *independent* activities, quite distinct in their nature. The repulsion process is accomplished exclusively only by two like charges. Mass is forever attractive to another mass. It can nowhere assume a repulsive activity.

Nevertheless, we may expect the unexpected in a different aspect, which concerns the *inseparability* between mass and charge. The independent activities of mass and charge inspire a smart fantasy that we must not entirely preclude a **separation** of a charge from its connate mass. In the long run, a way may be discovered which will probably enable this type of **separation**. If this comes true, then the force holding a charge to its connate mass must be extremely powerful, a **Marvelous force**, and must establish the utmost (highest) range of forces. This type of force will most probably be uncontrollable by man. It constitutes a *new type* of fundamental force, even more fundamental than either type of the two Basic Fields G & Em, not because of its superior power but because it is indeed a *new type* of force distinct from G and Em. It is not related to interactions between two masses or two charges; rather it involves *both* a mass and a charge in a yet unimaginable type of "gluing" factor.

Endnotes

1. Zamir, Y. pp. 7, 36 & 42. The Avkoan Theory adopted the word ELIXIR to suggest an eternal preservation of the basic attributes of matter. An Elixir is an intrinsic quality of matter that is unchangeable and unexchangeable with a corresponding Elixir from another particle unaccompanied by its connate Elixir, but not transformable into a different quality. Only an Avkoanic-Elixir can be an **ultimate unit** of mass or charge. Other Elixirs are **accumulative** units of several **Avkoanic-masses** or **Avkoanic-charges**, such as the mass of a planet and the charge of a metallic object. Consequently, only the Avkoanic Elixir is *indivisible*.
2. Pagels, H. *Perfect Symmetry.* p. 301
3. Zamir, Y. p. 19. The term **bi-by-product** is used here to suggest inseparable *twins* of opposite characteristics to each other, in contrast to positive and negative charges that are separable.
4. Zamir, Y. pp. 7 & 36. A **gender terminology of charges** is used in the Avkoan Theory instead of positive/negative charges to prevent a sense of destruction of particles/charges by neutralization. The Avkoan Theory disapproves the use of annihilation perception for the same reason.
5. Zamir, Y. p. 12.
6. Zamir, Y. p. 70.
7. Sullivan, W. *New York Times*, March 3, 1986

8. Pagels, H. *Perfect Symmetry*, Cf. p. 175 ff.
9. Hawking, S.W. *A Brief History of Time*, p. 70.
10. Einstein, A. et al. *The Principle of relativity*, p. 216.
11. Zamir, Y. p. 116 ff.
12. Pagels, H. *The Cosmic Code,* pp. 99-132
13. Weinberg, S.: The *First Three Minutes*, p. 94. ; Pagels, H.: *Perfect Symmetry*, p. 247
14. **A bare atom** is an atom that lost one or more electrons from its outer electron shell, or even from its inner shells. So, its nucleus becomes vulnerable to external attacks of negatively charged particles.
15. Atoms with two shells (Li, Be & B) are also excluded in case they give or share their outer shell electrons.
16. The proposed perceptions of heat and temperature may require revision of the laws of gases, which is out of our discussion in this limited topic of heat and temperature.
17. Redfield, J.: *The Celestine Prophecy*. Chapter 3 ff.

Bibliography

Einstein, A., *et al. The Principle of Relativity.* Dover: 1923

Hawking, S.W. *A Brief History of Time.* Bantam, NY: 1988

Pagels, H.R. *Perfect Symmetry.* Bantam, NY: 1986

_____. *The Cosmic Code.* Bantam, NY: 1984

Redfield, J. *The Celestine Prophecy,* Warner Books, NY.1994

Sullivan, W. *Theory of Fifth Force Spurs*...New York Times 3/3/86, NY: 1986

Weinberg, S.: The *First Three Minutes.* Bantam, NY: 1980

Zamir, Y. *Avkoan Theory, the Complete Volume I.* Los Angeles: 1994

About the Author:

Yecheskiel Zamir is the initiator of **The Principle of the Tiniest Particle** and the author of seven books expounding The **AVKOAN THEORY**.

Mr. Zamir graduated at the Hebrew University, Jerusalem, where he majored in Physics and minored in Math and Chemistry. During 20 years of teaching at Israeli Hi Schools, Mr. Zamir taught Physics, Math and Chemistry. He also supervised Hebrew teachers for new comer students. His visit to the United States in 1978 gave him a chance to teach in two Los Angeles schools, where he taught Hebrew, Math and Science. He retired in 1983 to complete his research in Particle Physics, which he started in the summer vacation of 1973. Recomposing his results into English was a challenge for him. The outcome is a publication of seven books expounding the Avkoan Theory.